James Edwin Duerden

Scientific Transactions of the Royal Dublin Society Vol. VI

XIV. Jamaican Actiniaria

James Edwin Duerden

Scientific Transactions of the Royal Dublin Society Vol. VI
XIV. Jamaican Actiniaria

ISBN/EAN: 9783743417397

Manufactured in Europe, USA, Canada, Australia, Japa

Cover: Foto ©berggeist007 / pixelio.de

Manufactured and distributed by brebook publishing software (www.brebook.com)

James Edwin Duerden

Scientific Transactions of the Royal Dublin Society Vol. VI

THE

SCIENTIFIC TRANSACTIONS

OF THE

ROYAL DUBLIN SOCIETY.

VOLUME VI.—(SERIES II.)

XIV.

JAMAICAN ACTINIARIA. Part I.—ZOANTHEÆ. By J. E. DUERDEN,

Assoc. R.C.Sc. (Lond.), Curator of the Museum of the Institute of Jamaica.

(Plates XVII. A, XVIII. A, XIX., XX.)

DUBLIN:
PUBLISHED BY THE ROYAL DUBLIN SOCIETY.

WILLIAMS AND NORGATE,
14, HENRIETTA STREET, COVENT GARDEN, LONDON;
20, SOUTH FREDERICK STREET, EDINBURGH; AND 7, BROAD STREET, OXFORD.

PRINTED AT THE UNIVERSITY PRESS, BY PONSONBY AND WELDRICK.

1898.

Price Three Shillings.

[April, 1898.]

QL
387
C704 ⌄.
19 0
TH

THE

SCIENTIFIC TRANSACTIONS

OF THE

ROYAL DUBLIN SOCIETY.

VOLUME VI.—(SERIES II.)

XIV.

JAMAICAN ACTINIARIA. Part I.—ZOANTHEÆ. By J. E. DUERDEN,

Assoc. R.C.Sc. (Lond.), Curator of the Museum of the Institute of Jamaica.

(Plates XVII. A, XVIII. A, XIX., XX.)

DUBLIN:
PUBLISHED BY THE ROYAL DUBLIN SOCIETY.
WILLIAMS AND NORGATE,
14, HENRIETTA STREET, COVENT GARDEN, LONDON;
20, SOUTH FREDERICK STREET, EDINBURGH; AND 7, BROAD STREET, OXFORD.
PRINTED AT THE UNIVERSITY PRESS, BY PONSONBY AND WELDRICK.

1898.

XIV.

JAMAICAN ACTINIARIA. PART I.—ZOANTHEÆ. By J. E. DUERDEN,

Assoc. R.C.Sc. (Lond.), Curator of the Museum of the Institute of Jamaica.

(PLATES XVII. A, XVIII. A, XIX., XX.)

[Read March 24, 1897.]

THE following account, restricted to the well-defined group of the Zoantheæ, is a first contribution from investigations now being carried out upon the Jamaican Actiniaria. It is remarkable that, with the exception of two species of Palythoa, collected by Sir Hans Sloane, probably about the year 1687, no Actinian has, so far as I can ascertain, been recorded from the island. Thanks to the labours of MM. Duchassaing and Michelotti (1850, 1860, 1866), and to the later researches of Professor M°Murrich (1889, 1889 a, 1896), we are acquainted with numerous examples from the other West Indian Islands, with which the Jamaican forms may be compared. These are proving that the Actinian fauna of the whole Caribbean region presents no marked difference. Professor M°Murrich has shown this for the Bahamas and the Bermudas, and of thirty-four Jamaican species now known, nearly all are forms recorded from one or more of the other islands of the Antilles. With the exception of the valuable work contributed by M°Murrich, practically no studies on Western forms have been conducted along the modern anatomical lines instituted and carried out elsewhere by Hertwig, Erdmann, Haddon, and others. Hence the necessity that the different representatives, many only partially known, should be submitted to microscopical examination to enable them to be arranged in the later systems of classification.

The following definition of the group of the Zoantheæ is the one given by Professor Haddon and Miss Shackleton (1891), and is practically the same as that accepted by all recent writers :—

ZOANTHEÆ.

Actiniæ with numerous perfect and imperfect mesenteries, and two pairs of directive mesenteries, of which the sulcar are perfect and the sulcular are imperfect. A pair of mesenteries occurs on each side of the sulcular directives, of which the sulcular moiety is perfect and its sulcar complement is imperfect; a similar second pair occurs in one section of the group (Brachycneminæ), or the second pair may be composed of two perfect mesenteries (Macrocneminæ). In the remaining pairs of mesenteries, of both divisions, this order is reversed, so that the perfect mesentery is sulcar and the imperfect is sulcular. The latter series of mesenteries are bilateral as regards the polyp, and arise independently (*i.e.* neither in pairs nor symmetrically on each side) in the exocœle on each side of the sulcar directives, in such a manner that the sulcular are the oldest and the sulcar the youngest. Only the perfect mesenteries are fertile or bear mesenterial filaments. A single sulcar œsophageal groove is present. The mesoglœa of the body-wall is traversed by irregularly branching octodermal canals or by scattered groups of cells. The body-wall is usually incrusted with foreign particles. The polyps are generally grouped in colonies connected by a cœnenchyme, the cœlenteron of each polyp communicating with that of the other members of the colony by means of basal endodermal canals.

Family. ZOANTHIDÆ, DANA, 1846.

(With the definition of the Group.)

Sub-family. BRACHYCNEMINÆ, Hadd. and Shackl., 1891.

Zoantheæ in which the sulcar element of the primitive sulco-lateral pair of mesenteries is imperfect.

GENERA.

Zoanthus, Lamarck, 1801.
Isaurus, Gray, 1828.
Gemmaria, Duchassaing and Michelotti, 1860.
Palythoa, Lamouroux, 1816.
Sphenopus, Steenstrup, 1856. (Not represented in the West Indian collections.)

Sub-family. MACROCNEMINÆ, Hadd. and Shackl., 1891.

Zoantheæ in which the sulcar element of the primitive sulco-lateral pair of mesenteries is perfect.

GENERA.

Epizoanthus, Gray, 1867.
Parazoanthus, Haddon and Shackleton, 1891.

One of the two type species of the genus Mammillifera, established by Lesueur, having been shown by McMurrich (1896) to possess all the characters of a Zoanthus, and there being every probability that the other type species, when discovered, will have the same, this genus, formerly included in the Zoanthidæ, can no longer be recognized (see p. 334).

With the exception of an irregular arrangement of the mesenteries in the genera Gemmaria and Palythoa, referred to below, nothing of importance has occurred differing from the diagnoses, mainly anatomical, of the tribe and genera given in the works of the writers referred to above.

Since the arrangement of the mesenteries in the Zoantheæ, which differs from that met with in all other Actiniaria, was first fully elucidated, it has also been recognized that the group presents a further distinction into two sub-divisions dependent upon the nature of the second pair of mesenteries on each side of the sulcular directives, according as the pair consists of a perfect and imperfect mesentery, or of two perfect mesenteries. The former was first termed by Dr. Erdmann the "microtypus," and the latter the "macrotypus"; a distinction emphasized later by Haddon and Shackleton (1891) in the formation of the two sub-families Brachycneminæ and Macrocneminæ. Apparently no variation from these two seemingly well-fixed divisions has since been noted. It is therefore interesting to find that, in West Indian species of the genus Gemmaria and of Palythoa, an irregular condition has been met with. In *Gemmaria variabilis*, n. sp., a specimen sectionized showed the normal brachycnemic arrangement on one side and the macrocnemic on the other. In a colony of *Palythoa mammillosa*, Ell. and Sol., one polyp was normally brachycnemic on the right, but macrocnemic on the left side; and in another polyp, in the same colony, the brachycnemic condition was on the left side and the macrocnemic on the right (Pl. xix. A, figs. 2 and 3). Similar combinations exist in *Palythoa caribæa*, Duch. and Mich.; but one polyp exhibited the full macrocnemic arrangement on both sides, in place of the normal brachycnemic (Pl. xix., fig. 7).

It is thus clear that in these three species, at least, the fundamental distinction of the microtype and macrotype is not sufficiently fixed, so that both may appear in one colony, or even on opposite sides in the same individual. In a number of specimens examined the majority are, however, normal.

In the present communication the following ten species are described:—

ZOANTHEÆ.

Brachycneminæ.

> **Zoanthus Solanderi**, Lesueur.
> **Zoanthus flos-marinus**, Duchassaing and Michelotti.
> **Zoanthus pulchellus** (Duchassaing and Michelotti).
> **Isaurus Duchassaingi** (Andres).
> **Gemmaria variabilis**, n. sp.
> **Gemmaria fusca**, n. sp.
> **Palythoa mammillosa** (Ellis and Solander).
> **Palythoa caribæa**, Duchassaing and Michelotti.

3 E 2

ZOANTHEÆ (*continued*).

Macrocneminæ.

Epizoanthus minutus, n. sp.

Parazoanthus Swiftii (Duchassaing and Michelotti).

All the examples having been partially studied in their living condition, and generally from an abundant supply of material, it has not been possible in some cases to draw up the specific characters in very hard and fast terms. The few external features one has to depend upon in the Zoanthidæ are well known to all workers in the group to be very variable; and especially will this be seen to be the case in the genera Zoanthus and Palythoa.

Recognizing the form of the sphincter muscle as of great importance in specific identification, I have figured it in all cases.

Practically all the material has been preserved by simple immersion in a four or five per cent. solution of formalin. Owing to the presence of abundant incrustations and the great thickness of the mesoglœa, the internal tissues of the Zoanthidæ are rarely well preserved. It is satisfactory to find that, by means of formalin, the preservation and histology was, in every case, all that could be desired, while, in most, little alteration of form or dimensions occurred; thus allowing the number of capitular ridges and tentacles, measurements, etc., to be taken at leisure. The colours can likewise be observed for some time, but disappear ultimately.

A curious chromatic change occurred in most of the Palythoa material. The colonies, usually cream colour when alive, became strongly brick-red in their upper region after immersion for a short time in the formalin. This alteration extended also to the ectoderm of the œsophagus and to the mesenterial filaments.

Some importance must be attached to the method of preservation in determining the appearance of the various histological characters. With alcohol the mesoglœa shrinks very considerably. To this disproportionate shrinkage, compared with that of the ectoderm and cuticle, is due the contorted or dendriform appearance of the outer part of the body-wall often seen in species of Zoanthus. The size and appearance of the mesoglœal cavities, especially those containing the sphincter muscle, may be much modified. Figs. 1, 2, and 3, on Pl. xviii. a, should be compared. The last having narrow, almost closed cavities, was drawn from a polyp shrunk in alcohol, and the two first from colonies preserved without shrinkage in formalin. Specimens of *Z. pulchellus* preserved later in formalin show open cavities like figs. 1 and 2. The figure of the cavities of the sphincter muscle of *Zoanthus*, sp. ?, given by Hertwig (1882, pl. xiv., fig. 1), is evidently partly determined by this shrinkage.

ZOANTHUS, LAMARCK, 1801.

Brachycnemic Zoantheæ, with a double mesoglœal sphincter muscle. The body-wall is unincrusted; the ectoderm is usually discontinuous; well-developed ectodermal canal system in the mesoglœa. Monœcious or diœcious. Polyps connected by a thin lamellar cœnenchyme, stolons, or, more rarely, free.

The synonymy of the genus Zoanthus is given by most recent writers upon the Zoanthidæ.

The following are the references to the genus Mammillifera, which, as shown below, must be merged in Zoanthus:

Mammillifera,	Lesueur, 1817, p. 178.
Mamillifera, .	Blainville, 1834, p. 329.
Mammillifera,	Ehrenberg, 1834, p. 46.
Mamillifera, .	Duchassaing, 1850, p. 11.
Palythoa (pars.)	Milne-Edwards, 1857, p. 301.
Zoanthus, .	Gosse, 1860, p. 296.
Mamillifera, .	Duchassaing and Michelotti, 1860, p. 327; 1866, p. 136.
Mammillifera,	Verrill, 1869, p. 495.
Mammilifera, . . .	Hertwig, 1882, p. 111.
Polythoa and *Zoanthus* (pars.),	. Andres, 1883, pp. 306 and 323.
Mammillifera, , , ,	. Erdmann, 1885.
Mammilifera, . .	Hertwig, 1888, p. 35.
Mammillifera,	McMurrich, 1889, p. 117.
Mammillifera,	. Haddon and Shackleton, 1891, p. 630.
Zoanthus, .	. McMurrich, 1890, p. 188.

The distinction between the Zoantheæ and other Actiniæ was first recognised by Cuvier in 1798, and the genus Zoantha first employed by Lamarck in 1801. Later, Cuvier (1817) restricted Zoanthus for Actinians occurring in groups adhering to a common base, which is sometimes broad and flat, and at other times a sort of creeping stem. In the same year, Lesueur (1817) separated, under the genus Mammillifera, those which have " A large cuticular expansion, serving as the base of numerous animals which, when contracted, assume the form of mammæ." This genus was received by Blainville (1834), Ehrenberg (1834), and Duchassaing (1850). Milne-Edwards (1857) united the included species under the genus Palythoa, established by Lamouroux (1816). Duchassaing and Michelotti (1860 and 1866) again separated the two genera, restricting Palythoa to the forms which have the integuments hardened by incrusting foreign matter. Gosse (1860) embraced, under Zoanthus, both the Palythoa of Lamouroux and the Mammillifera of Lesueur in addition to the other genera, all forms spreading "in either a linear or incrusting manner." Verrill (1869) distinguished the genus Mammillifera from Zoanthus "in having smaller, shorter, or more sessile polyps, and in the tendency to form continuous basal membranes, instead of linear stolons." Hertwig (1882) states that "Zoanthus and Epizoanthus are distinguished from Mammilifera and Palythoa by the fact that, in the former two, the polyps project plainly above the common basis, whilst in the latter two they are united up to the free end by basal cœnenchyma." Andres (1883) distributes the various species partly under Polythoa and partly under Zoanthus. Hertwig (1888), as a result of the researches of Erdmann (1885), separates Mammillifera from Zoanthus by the possession of only a simple mesodermal sphincter muscle. McMurrich (1889) follows Erdmann in this. Haddon and Shackleton, however, in their "Revision of the British Actiniæ" (1891), in a footnote to Mammillifera, state :—"The position of this genus cannot be settled until the type species have been recovered and sectionized"—the types being Lesueur's *Mammillifera auricula* and *M. nymphæa*. A form, covering considerable areas, occurs at Port Henderson and at Drunkenman Cay,* near Kingston, which I have no hesitation in identifying as closely allied to the *M. nymphæa*, of Lesueur, and with the better description of the species given by Duchassaing and Michelotti for colonies found by them at different islands of the West Indies. An anatomical study of this shows that, not only in the fundamental characters of the brachycnemic arrangement of the mesenteries and the double mesoglœal sphincter muscle, but also in many minuter details of structure, the species agrees with other members of the genus Zoanthus as described by Erdmann, McMurrich, and Haddon. Lesueur's description and figure of *M. auricula*, the other type species, leave no doubt that, when found and examined, it will also have the characters of a Zoanthus.

Since this was first written, Prof. McMurrich (1896) has obtained from the Bahamas an incrusting form which he identifies as *Mammillifera nymphæa*, and has shown that it is an undoubted Zoanthus. It is distinct from the Jamaican species (p. 345).

Following Haddon, McMurrich removes the species placed in his earlier paper under Mammillifera to Isaurus.

The use of the sphincter muscle, for specific purposes, is well exemplified in the genus Zoanthus. Differences are readily seen in the figures of the three following species, and these again can be distinguished from the sphincters of others represented elsewhere. A marked difference, in the muscle, exists in

---

* The Port Royal Cays, known as Gun, Rackum, Drunkenman, Lime, Maiden, South, and South-east Cays, are a group of small coral islands outside Kingston Harbour. They are raised but a little above sea-level, some with and others without vegetation. The shores and shallow-waters around are the usual and most favourable spots for marine collectors.

the three Torres Straits species, described by Haddon and Shackleton (1891), compared with those of the West Indies. In all these latter, the proportion of the two parts differs much, the lower or proximal being several times larger than the upper or distal; in the former, very little distinction in size is met with. In *Z. Coppingeri*, the proximal (*i.e.* upper in figure, which would be lower or proximal in the extended condition of the polyp) is even slightly shorter than the distal; while in *Z. Jukesii* and *Z. Macgillivrayi*, the proximal is but slightly longer. Further, the muscle as a whole is much less developed than in the Antillean examples.

In partial contraction, a deep circular depression denotes externally the place of division between the two portions of the sphincter muscle. I use the term 'capitular fossa' for this, and speak of the two parts as the inner and outer capitula.

The genus Zoanthus, so far as I have observed it in the abundance occurring in Kingston Harbour and the Port Royal Cays, appears restricted in its distribution to a narrow belt of shallow water around the shores. It contrasts very markedly, not only in colour and firmness, but in its distribution with the equally abundant genus Palythoa. Colonies of the latter commence where the former begins to disappear, being most vigorous in the upper region of the breakers around the reefs. On the windward, more rocky side of the Cays, a distinct Zoanthus zone can be distinguished from the Palythoa zone; the former extends to a depth of one or two fathoms, and the latter to three or four fathoms, and gives place in its turn to the zone of living coral.

Zoanthus Solanderi, Lesueur.

(Pl. xvii. a, fig. 1.)

Zoanthus Solandri,	Lesueur, 1817, p. 177, pl. viii., fig. 1.
Zoanthus dubia, .	. Lesueur, 1817, p. 177.
Zoanthus Solanderi,	. Milne-Edwards, 1857, p. 300, pl. c 2, fig. 3.
Zoanthus dubius, .	. Milne-Edwards, 1857, p. 300.
Zoanthus Solanderi,	. Duchassaing and Michelotti, 1860, p. 325, pl. viii., fig. 1 ; 1866, p. 135.
Zoanthus dubius, .	. Duchassaing and Michelotti, 1860, p. 326, pl. viii., fig. 2 ; 1866, p. 135.
Zoanthus (Rhyzanthus) Solanderii,	Andres, 1883, p. 327.
Zoanthus (Rhyzanthus) dubius,	. Andres, 1883, p. 329.

Form.—Polyps erect, cylindrical, smooth, thin-walled with lines of attachment of mesenteries showing through, connected with one another at the base by lamellar narrow cœnenchyme or free stolons, or may be solitary. Column usually non-pedunculate, practically of the same diameter throughout, but often with slightly expanded portions at the base. In extension the margin of the column is crenate, the elevations alternating with the outer row of tentacles; on partial retraction, the capitular fossa is well seen. Both inner and outer capitula bear fine ridges and grooves. In complete retraction, numerous minute capitular striæ can be seen only on the outer capitulum. Tentacles about 60, arranged in two cycles; one specimen had 32 in each row. Disc not much depressed, walls thin; the mesenterial lines can be seen through them ; mouth slit-like, the œsophageal groove not distinguishable ; œsophageal walls thin, showing the mesenterial lines; in some examples, the wall is thrown into ridges and furrows. Cœnenchyme little developed, appears only as a flattened expansion from one side of a polyp and connected with one or more other polyps near; in other cases, the connecting strand becomes constricted and stolon-like. Isolated polyps devoid of any cœnenchyme are met with, even when closely associated.

The polyps and cœnenchyme adhere firmly to the rocks or stones. New individuals arise by budding from the slightly expanded base of other polyps, and afterwards become more separated, the connecting tissue getting thinner and thinner until the polyps may become entirely isolated.

Colour.—Column in its lower part, especially when embedded in foreign matter, sand-coloured, becoming a dark blue or slate colour above. The margin has irregularly disposed, silvery white, triangular, narrow, radiating patches, often incomplete and variable in length ; the toothed elevations are nearly opaque white. These white markings, which appear constant for the species, are best seen on partial contraction, at which time the inner capitulum appears as a distinct whitish, toothed, circular annulus. Colours of the tentacles and disc are variable. In a colony from Lime Cay both were a bright orange brown, and the peristome a bright green ; in another large colony from Maiden Cay the tentacles were green on their inner aspect and dark brown on the outer, the disc a dark brown with bright green peristome ; specimens at Rackum Cay showed a bright blue disc with green lips, and the tentacles a bright green.

Dimensions.—The height differs considerably, dependent upon the position of the polyps in a colony; it may vary from 2·7 cm. to 0·4 cm.; the diameter is about 0·6 cm., and is generally constant throughout the column, and independent of the length. The measurements are taken from specimens preserved in formalin, with but little contraction.

Locality.—Jamaica: Found in considerable abundance, growing on stones and coral rock, in shallow water, around the various Cays outside Kingston Harbour. The polyps are often partially embedded in sand and shore débris.

Range.—St. Thomas, Guadaloupe (Lesueur; Duchassaing and Michelotti).

Column-wall (Pl. XVIII. A, fig. 1).—A cuticle and sub-cuticla are present, the former having much adhering matter, such as diatoms; the sub-cuticla is in places minutely convoluted, evidently as a result of the excessive shrinkage of the mesoglœa. This is more especially seen in longitudinal sections.

The ectoderm of the column-wall is broad above, but very narrow below. It is much vacuolated, with only a small amount of cellular tissue remaining in the form of strands passing from the outer to the inner boundary; rarely a connecting strand of mesoglœa is seen. Some examples are not so highly vacuolated, particularly in the lower part. Abundant medium-sized, oval, non-staining nematocysts are present, the inner thread showing distinctly; pigment granules are met with at its internal boundary.

An irregular layer of spherical lacunæ appears in the mesoglœa, immediately below the ectoderm. Proximally the empty spaces extend further into the meso-glœa. The mesoglœa is broad in the region of the sphincter muscle, but narrows much below. Cells with long, fine processes are distributed sparingly throughout; delicate fibrils can also be easily seen passing from the ectoderm to an irregular, much broken, encircling sinus. The latter, situated either about the middle or very near the endodermal border of the mesoglœa, is formed of spaces varying in dimensions and form. They contain a small quantity of cellular tissue, and are connected with one another by larger or smaller canals, and also by canals with the ectoderm and the endoderm. The cells are multipolar in character.

The endoderm is low and contains abundant zooxanthellæ; nematocysts, similar to those in the ectoderm are present, and a weak circular muscle.

Sphincter muscle (Pl. XVIII. A, fig. 1).—The sphincter muscle is mesoglœal and double, the two halves being distinctly separated. The upper or distal is smaller, and located in large irregular cavities extending almost across the mesoglœa, diminishing both proximally and distally; the lining of muscle cells is thin. The lower or proximal part of the muscle is contained in a large number of small, scattered, mostly circular, mesoglœal cavities; the distal ones, however, are elongated, more like those in the upper. The muscle fibres are very small in section, and only a little loose tissue is present in addition.

Tentacles.—The ectoderm of the tentacles is without cuticle or sub-cuticla, and shows two kinds of nematocysts—an outer thick zone of the usual narrow form, and an occasional medium-sized, oval-shaped form similar to those in the ectoderm of the column-wall. The mesoglœa is very thin.

The endoderm is well developed, and crowded with zooxanthellæ. Both the ectodermal and endodermal musculatures are weak.

Disc.—The ectoderm of the disc is almost devoid of nematocysts. The mesoglœa is a little thicker, and the endoderm much thinner than in the tentacles. The endodermal muscle is seen in longitudinal sections.

Œsophagus.—The nematocyst and nuclear zone in the ectoderm is very regular, and situated close to the outer surface, while a non-staining nervous tissue intervenes between it and the mesoglœa. The latter is very thin, and the endoderm resembles that of the mesenteries. In transverse sections, the œsophagus is oval. In the upper region, the ectoderm is not thrown into folds, and the œsophageal groove is barely apparent. Lower, as many as twelve longitudinal folds may be present on each side, and a slight indication of a groove. The mesenteries are attached to the œsophagus at about equal distances all the way round. The ectoderm is reflected on the mesenteries, and continued downwards as the mesenterial filaments.

Mesenteries.—The mesenteries are of the brachycnemic type ; generally about thirty pairs are present ; one specimen had fifteen perfect mesenteries on one side, and only thirteen on the other.

The endoderm contains abundant zooxanthellæ, and medium-sized oval nematocysts. The digestive endoderm (1889, p. 116 ; 1891, p. 622) is not very thick.

The basal canal is large in both the perfect and imperfect mesenteries. It is elongated in the former, and full of deeply-staining cells.

The parieto-basilar muscles are clearly distinguishable, as also the retractor muscle of the mesenteries. The mesoglœa is folded to support the muscle.

Gonads.—No reproductive elements were present in any of the examples studied.

The following may be regarded as distinguishing anatomical characters:

(*a*). Pigment limited to the inner portion of ectoderm ;

(*b*). Mesoglœal lacunæ ;

(*c*). Form of sphincter muscle.

The Jamaican form above described appears to unite the two species *Z. Solandri* and *Z. dubia*, as originally described by Lesueur, and as known to Duchassaing and Michelotti from the same localities. Considering the variation in colour noted, it is evident that little importance can be attached to it. Referring to the first species, Duchassaing and Michelotti state : " Le couleur de cette espèce est sujet à varier, mais nous n'y avons jamais remarqué à l'état vivant la teinte qui lui donne M. Milne-Edwards dans l'atlas qui accompagne son oûvrage sur les coralliaires, la teinte verte se montre toujours dans une parti ou sur la totalité du corps de ces animaux." Some account must be taken of the character given by Lesueur that, when the animal is contracted, the summit is marked with deep blue angular spots and white lines, a feature agreeing with the present specimens. The polyps are usually non-pedunculate, resembling the figure of *Z. dubius* given by Duchassaing and Michelotti ; but pedunculate forms, agreeing with Lesueur's original figure, also occur.

Zoanthus flos-marinus, Duchassaing and Michelotti.

(Pl. xvii. a, fig. 2.)

Zoanthus flos-marinus,	. Duchassaing and Michelotti, 1860, p. 326, pl. viii., fig. 6.
Zoanthus flos-marinus,	Andres, 1883, p. 328.
Zoanthus flos-marinus,	. McMurrich, 1889, p. 113, pl. vii., figs. 3, 4.

Form.—Polyps erect, smooth, thin-walled, pellucid, clavate or cylindrical; arising either directly from a thin band-like incrusting cœnenchyme, or from a free irregular stolon, or from the base of one another. In full retraction, a little swollen above; in partial contraction, inner capitulum very narrow, with 24 to 30 minute rounded denticulations or capitular ridges, continued as thin lines for some distance down the column, and corresponding in number and alternating with the outer row of tentacles.

Tentacles dicyclic, slightly entacmæous, smooth, acuminate, overhanging in full extension, variable in number, from 48 to 60. In one colony, the numbers counted were 60, 52, 58, 54, 50, 54, 58; in another colony, 56, 50, 50, 48. In this latter colony, a curious condition of the tentacles was met with, each bearing near its origin one or two small tubercles,* suggestive of an additional cycle.

Disc thin-walled, with the radiating mesenterial lines showing through; outer part grooved, overhanging in full extension, central portion elevated and rounded; mouth slit-like. Cœnenchyme occasionally band-like and incrusting, more often stolon-like, constituting an irregular connexion for the polyps. Polyps, all about the same size, are often closely associated in a colony, and incrust some rock or stone; at other times, they are loosely attached to any object, and form bunches connected with one another in an irregular fashion by the loose stolon-like cœnenchyme. Sometimes the polyps are united to one another some distance above the base. Examples on the upper surface of stones are usually short and cylindrical; but those along the sides and underneath, or in crevices, become much elongated and narrow below.

Colour.—Lower part of column sand-coloured; upper dark green or lead colour; tentacles yellowish-green, blue-green, or brown; disc various light and dark shades of blue and green, often mixed with yellow and black; peristome a bright yellow or green; a darker triangular area at each or only one angle of the mouth may be present.

Dimensions.—Dimensions variable; column usually about 1·7 cm. in length;

---

* Verrill records a similar condition for *Mammillifera Danæ* (1869, p. 496), and for *Epizoanthus elongatus* (p. 498). It is not general in the present species.

diameter of capitulum in living retracted state 0·5 cm.; diameter of disc in extension 0·5 to 0·8 cm.; inner tentacles 0·25 cm. long.

Locality.—Jamaica: The commonest Zoanthus found around all the Cays. It occurs in masses, covering large surfaces of the rocks and stones in shallow water. Very often the polyps are almost embedded in débris of sand, mud, and calcareous algæ, so that in extension only the closely associated discs are exposed.

Range.—Bermudas (MᶜMurrich); St. Thomas (Duchassaing and Michelotti).

Column-wall.—The cuticle, sub-cuticle, and ectoderm are of the same character as in the previous species. In preserved specimens the cuticle readily separates. Abundant oval nematocysts are present in the ectoderm, especially in the distal part. The boundary between the ectoderm and mesoglœa is not well defined, cells and cell processes from the former passing into the latter.

The mesoglœa is broad in the region of the lower sphincter muscle, but becomes thinner in both directions. It is without the empty lacunæ below the ectoderm, which are such a marked feature in the former; large and small spaces occur, the former containing but little cellular tissue and an occasional nematocyst. In transverse sections a broken encircling canal is shown, in some sections communicating with the ectoderm. Most of the cell-islets throughout the mesoglœa contain fine pigment granules. The endoderm is occasionally elevated between the mesenteries, and triangular in transverse sections; elsewhere it is very thin, and loaded with zooxanthellæ. The endodermal muscle is clearly distinguishable.

Sphincter muscle (Pl. XVIII. A, fig. 2).—The form and arrangement of the cavities of the sphincter muscle are best realized from the figure. It bears a resemblance to the previous one, but the smaller proximal cavities are much more uniformly and regularly distributed. The smaller cavities terminating the proximal half are more numerous in some examples than in the one figured. In addition to the lining muscle cells, rounded cells occur in the cavities.

Tentacles.—The ectoderm is devoid of the cuticle and sub-cuticle. It is made up of narrow columnar cells, with oval, deeply-staining nuclei, amongst which are small oval nematocysts; pigment granules and a weak ectodermal muscle occur, the latter on very numerous, fine, mesoglœal plaitings. The mesoglœa is thin and a little plaited on the endodermal border for the support of the circular muscle.

The endoderm is very thick, leaving only a small lumen; it is crowded with zooxanthellæ.

Disc.—The disc is much like the tentacles in structure, but the endoderm has about the same thickness as the ectoderm.

Œsophagus.—In section, the ectoderm of the œsophagus shows three strongly marked zones, all of nearly equal breadth; an outer non-staining ciliated portion; a middle deeply-staining zone with oval-shaped nuclei, granular gland cells and narrow nematocysts; and an inner, slightly narrower, nervous layer, containing a few circular nuclei, and a little pigment matter. The mesoglœa and endoderm are each narrow. In transverse section, the œsophageal groove is not very pronounced, and the ectoderm is thrown into folds in some cases, in others not.

Mesenteries.—The mesenteries are brachycnemic in arrangement and very thin. In one specimen, twenty-four pairs were present; in another, twenty-one. The endoderm is well developed, made up almost entirely of zooxanthellæ and medium-sized, oval-shaped nematocysts. The mesoglœa is folded and plaited on one side for the support of the longitudinal retractor muscle. A basal canal is developed in some a little distance from the column-wall, but is not present in others.

The reflected ectoderm, mesenterial filaments, and endoderm swollen in the lower region, are similar to those figured and described by McMurrich (1889, p. 115, pl. vii., figs. 3, 4), and the endoderm has embedded in it what I take to be the delicate acicular siliceous spicules referred to by him. The digestive endoderm is not so thickly developed in the previous species. The Drüsenwulst of von Heider (1895, p. 129) can be well studied.

Gonads.—None of the numerous specimens examined were fertile.

I identify this very common Jamaican form as the *Zoanthus flos-marinus* of Duchassaing and Michelotti, rather from the description by Prof. McMurrich of specimens from the Bermudas (1889). The diagnosis of the original authors is very incomplete for this variable genus. They state the tentacles to be thirty-six, while the later writer gives them as fifty to sixty in number, a number agreeing with the Jamaican examples. Andres places it amongst his *Zoanthi dubii.*

It may readily be distinguished from *Z. Solanderi* by its smaller size, usually clavate form, and stolon-like cœnenchyme; and from *Z. pulchellus* by never forming a broad lamellar cœnenchyme.

Zoanthus pulchellus (Duchassaing and Michelotti).

(Pl. XVII. A, fig. 3.)

Mamillifera pulchella, Duchassaing and Michelotti, 1866, p. 137, pl. vi., fig. 4.
Polythoa (Mammothoa) nymphosa, Andres, 1883, p. 320.

Form.—Polyps erect, cylindrical, short or elongated, smooth, usually closely grouped, rising from a thin, tough, lamellar, incrusting cœnenchyme. In retraction, either a little enlarged above or of the same diameter throughout, terminating in a rounded or slightly conical manner, and showing a central aperture and numerous fine radiating capitular ridges; where the polyps are more separated, they often appear as low mammiform prominences. In partial retraction, a double capitulum is formed by the groove situated between the two parts of the sphincter muscle. In full expansion, the disc and capitulum are greatly extended, so that, when all the polyps in a colony are in this state, their margins are wholly in contact. The mutual pressure produces a polygonal outline, giving rise to the appearance of a mosaic work of green discs with elevated, often pink, centres, the two rows of dark short tentacles simulating a thick cementing material. Tentacles short, digitiform, overhanging in extension, arranged in two alternating rows of about thirty in each. The number may be slightly more or less. Disc depressed below the thickened margin; the mesenterial lines are seen through the wall; in expansion the oral cone is considerably elevated, and the mouth slit-like; the œsophageal groove is not obvious. The cœnenchyme is smooth, continuous, lamellar, adhering firmly to the rocks and stones, and following the larger irregularities of the surfaces. The polyps all arise independently, generally in close association, but may be further separated, when the cœnenchyme becomes more ribbon-shaped. Owing to the thinness of the body-wall, there is often a partial collapse and transverse wrinkling in alcoholic specimens, especially in the more elongated examples.

Colour.—Column in lower part of elongated forms is pale buff and transparent, with the white mesenterial lines showing through; upper part olive blue; capitulum lighter with green radiating lines, seen more especially on retraction. Tentacles, nearly always dark brown, may be green or olive. Disc generally a bright green, with light radiating lines corresponding with the internal mesenteries; sometimes a pale green or yellow. In many, a darker triangular area extends towards the margin from each of the two extremities of the mouth; one is often more pronounced than the other. Œsophagus green, with white lines showing through. Peristome in many colonies pink, in others a bright green; more rarely yellow. An olive brown colour is first extracted by alcohol, leaving the colonies uniformly dark green, probably due to the abundant internal zooxanthellæ; later the polyps become a buff colour, a little darker above, and the mesenterial lines show through.

Dimensions.—Average diameter of column, 0·6 cm.; diameter of capitulum, in full expansion, 0·8 to 1 cm.; length of column very variable, depending largely upon the position of the polyp in the colony, average length 1·3 cm.; some may attain a length of nearly 3 cm., while others extend only 0·4 cm. above the cœnenchyme. Tentacles 0·2 to 0·3 cm. in length. Colonies often 20 or 30 cm. across. When preserved in alcohol, considerable contraction of the polyps occurs.

Locality.—Jamaica: Found in great abundance, forming large incrusting colonies on the rocks and stones in the shallow waters near the rocky parts of the shore at Port Henderson, Kingston Harbour, and on the coral-rock at Drunkenman and other Cays.

Range.—St. Thomas (Duchassaing and Michelotti).

Column-wall (Pl. XVIII. A, fig. 3).—The column is partially coated with a layer of foreign matter, mostly diatom frustules and fine mud. The ectoderm is very thin, nearly continuous, and only slightly vacuolated. A sub-cuticla occurs, as in most species of the genus, more noticeable on the lower part; transverse strands of mesogloea are rarely seen. The mesogloea is very variable in thickness, according to the state of extension or contraction of the polyp; it is best developed in the region of the sphincter muscle, and also as the coenenchyme is approached; isolated cells, with elongated processes, occur; fine processes are seen extending across the mesogloea from the ectoderm to the endoderm. Some of the more peripheral cell-islets contain dark granular pigment matter. An irregular, partially encircling, canal system is present, situated in the upper part nearer the endoderm, among the cells of which are nematocysts. The canals, in some sections, are seen definitely connected with the ectoderm.* The mesogloea is much shrunk in preserved specimens, producing, especially in longitudinal sections, a very irregular external outline, followed by the ectoderm and foreign material. The endoderm is very narrow, crowded with zooxanthellæ and small oval nematocysts, and gives rise to a weak endodermal muscle.

At the base, the ectoderm is thinner; the sub-cuticla is more clearly seen, also the ectodermal canals in the mesogloea communicating with the ectoderm. Numerous irregularly distributed coelenteric canals, lined with ciliated epithelium, pass along the base of the polyps through the coenenchyme, and connect the cavity of one polyp with that of another. The cells of the canals are somewhat glandular, and a thin lining musculature is present.

Sphincter muscle (Pl. XVIII. A, fig. 3).—The upper (distal) portion of the sphincter muscle is much smaller than the lower (proximal). It is contained in about twenty small mesogloeal cavities, arranged in an irregular row. The first section of the larger muscle is contained in an irregular series of small cavities stretching for some distance across the mesogloea. The cavities are largest about the middle; lower they are again smaller, and located for the most part nearer the ectoderm. The lining of muscle cells is very thin, a few nucleated rounded cells are also present. In the figure of the muscle cavities, the latter are represented as flattened and almost closed. This condition is evidently due to the method of preservation in alcohol. Specimens preserved later, in formalin, have

---

* Hertwig (1882, p. 112) found a similar connexion in *Zoanthus Danæ* (?). McMurrich states that, in *Z. sociatus*, he has observed the basal canal in the mesentery communicating with one of these spaces, and considers it open to question whether the cells in the large cavities of the mesogloea are not in reality endodermal in their origin.

344 J. E. Duerden—*Jamaican Actiniaria: Part I.—Zoantheæ.*

the cavities larger and more circular, as in the figures of the two previous species. The sphincter in this species differs from that of the two former in that the proximal part commences above with numerous small cavities.

Tentacles.—The ectoderm of the tentacles is ciliated and without any cuticle or sub-cuticla; it is much thicker than that of the column, and small nematocysts occur in restricted areas. The weak ectodermal muscle is supported on minute plaitings of the mesoglœa. The mesoglœa is thin, with a few isolated cells. The endoderm is thicker than the ectoderm, leaving scarcely any lumen in retraction. Abundant zooxanthellæ, small oval nematocysts, and a weak endodermal muscle are met with.

Disc.—The ectoderm of the disc is nearly as thick as that of the tentacles ; the nuclei stain very deeply ; an ectodermal musculature occurs. The mesoglœa is broad, destitute of cell-enclosures, and contains a few isolated cells. The endoderm is like that of the mesenteries, and has an endodermal musculature.

Œsophagus.—The ectoderm in the œsophagus is rather broad and ciliated, and thrown into about eight deep longitudinal folds on each side, partially followed by the mesoglœa ; the œsophageal groove is elongated, occupying in some sections about one-third the transverse area of the œsophagus. The appearance is much the same as that figured by McMurrich for the stomadœum of *Z. sociatus.* The ectoderm contains an occasional small oval nematocyst, in addition to the usual abundant narrow ones, and also a little pigment matter on its inner border. The mesoglœa is thinner than the ectoderm, and contains no cell enclosures.

Mesenteries (Pl. XVIII. A, fig. 4).—The number of mesenteries varies, twenty. eight perfect ones occurring in one specimen, and twenty-six in another, corresponding with the varying number of tentacles. A slight parieto-basilar muscle is found on each side. The endoderm has zooxanthellæ and small nematocysts. A few isolated cells occur in the mesoglœa. Below the œsophagus, the mesenteries, with the mesenterial filaments, assume, in transverse section, first a sagittate appearance, and lower a clavate form ; in the lower region of the œsophagus, the reflected ectoderm has the characteristic pinnate appearance, the whole corresponding with that described and figured by Haddon and Shackleton for *Z. Macgillivrayi* (1891, p. 681). Nematocysts occur.

A very weak musculature extends along the whole surface of both sides of the mesentery, the mesoglœa being slightly plaited in places. A single basal canal passes the whole vertical length ; in the perfect mesenteries, it is oval in section in the distal region, but becomes elongated and stretches nearly the whole width in the basal part of the polyp.

In the imperfect mesenteries, the basal canal remains approximately circular in section. The tissue inside the canals is of the same character as the endoderm, being crowded with zooxanthellæ and nematocysts.

Gonads (Pl. XVIII. A, fig. 4).—Ova and spermaria are borne in close proximity on the perfect mesenteries, both above and below the œsophagus. They were met with in three specimens taken from the same colony, but none were present in several examples sectionized from another colony.

Cœnenchyme.—The cœnenchyme is of similar structure to the body-wall, but the mesogloea is much thicker and broken up by large ciliated cœlenteric canals passing in all directions; the endodermal lining is loaded with zooxanthellæ, and has a weak musculature. Isolated cells with fine processes, and the smaller ectodermal canals occur.

For some time during the preparation of this Paper, I had regarded this species, with little or no hesitation, as the *Mammillifera nymphœa* of Lesueur (1817). In the meantime, Prof. M°Murrich identified, with some amount of uncertainty, a form from the Bahamas as Lesueur's species. The external characters of the Jamaican representative agree with those of the Bahaman, as far as the latter are given, but a comparison of the sphincter muscles shows that they are undoubtedly distinct. Prof. M°Murrich, from an examination of my material and slides, entirely agrees with this. Whether his identification of the Bahaman form with that which Lesueur described be correct or not, it seems best that his conclusion should be followed for the future, seeing that with the addition of the anatomical features, the characters of the species are definitely fixed once for all. There must nearly always be an amount of uncertainty in identifying the species of the older authors, where external characters only were taken into account. I have therefore changed my manuscript identification of this species to that of *Mammillifera pulchella* of Duchassaing and Michelotti (1866), a form these authors regarded as a doubtful variety of *M. nymphœa*.

Isaurus, GRAY, 1828.

Large brachycnemic Zoanthew, with a single mesoglœal sphincter muscle. The body-wall is unincrusted; the ectoderm discontinuous; ectodermal and endodermal bays and small canals in the mesogloea. Monœcious or diœcious. Polyps in small clusters or solitary.

Prof. Haddon and Miss Shackleton give (1891, pp. 682–4) a full discussion of the genus defined by them as above. They dwell particularly upon the reason why it should not be merged into the genus *Mammillifera* of Lesueur, as, accepting the characters Erdmann gives (1888, p. 35), has been done by M°Murrich (1889, p. 117). In his more recent paper (1896, p. 191), the latter author adopts *Isaurus*.

Isaurus Duchassaingi (Andres).

(Pl. XVII. A, fig. 4.)

Zoanthus tuberculatus,	. Duchassaing, 1850, p. 11.
Zoanthus tuberculatus,	. Duchassaing and Michelotti, 1860, p. 327, pl. viii., fig. 5.
Antinedia tuberculata,	. Duchassaing and Michelotti, 1866, p. 136, pl. vi., figs. 2, 3.
Antinedia Duchassaingi, .	. Andres, 1883, p. 330.
Isaurus Duchassaingi, .	. Mᶜ Murrich, 1896, p. 190, pl. xvii., figs. 6–8.

Form.—Base firmly adherent, expanding somewhat over the incrusted surface; usually much larger than the diameter of the column; irregular in outline. The flattened expansion may be regarded as a slightly developed cœnenchyme; but, although closely associated, the polyps were rarely connected with one another.

Column variable in shape, may be cylindrical or clavate, generally more expanded towards the base; slightly overhanging so as to present a concave and a convex aspect; the capitulum appears as a disc in retracted specimens, and is placed obliquely so that the small central aperture indicating the mouth is below the upper termination of the column. Proximal part, for from one-third to one-half of the total length of the column, smooth, with thin partial annuli showing through, which may become depressions on shrinkage; in the later state, numerous well-marked longitudinal ridges and furrows may also be rendered obvious. The column on its sides and convex aspect bears irregular rows of rather large, rounded tubercles, distinct from one another; the concave, shorter portion is smooth, giving rise to a marked asymmetry; four principal rows, of from five to eight tubercles, alternate with other rows of two or three smaller protuberances. Around the margin of the terminal disc is an incomplete circle of eight or nine tubercles, separated by deep depressions. These correspond with the rows and extend nearly round the margin, diminishing in size towards each extremity of the partial circle.

The flattened or slightly elevated and dome-shaped capitular disc is partially enclosed by these, and bears radiating ridges and furrows, not all equally developed; eight or nine which alternate with the marginal protuberances are more prominent than the one or two groups alternating with them. A depression indicates the position of the mouth. One young specimen, 2·5 cm. long, is quite smooth, having no elevations. In the living condition, the column-wall is firm, very tough, and partially transparent, so that the presence of the internal organs can be distinguished.

In none of the specimens could the disc and tentacles be noted externally ; the individuals, as appears to be usually the case, maintaining a retracted condition.

Colour.—Cœnenchymatous base colourless in some, irregularly greenish brown in others ; column dark brown, mottled with green and black, the pigment appearing in granular form. The green colouring matter seems largely external, and due to adhering unicellular algæ.

Dimensions.—Diameter of base may be 1·7 cm. ; diameter at commencement of column varies from 0·4 cm. to 1 cm. ; average diameter of column 0·6 cm. ; height, from 2·2 cm. to 4·2 cm. ; tentacles, measured in sections, 0·3 cm. long.

Locality.—Seven specimens were found associated and firmly adhering to a small block of coral-rock on the south-east side of Drunkenman Cay.

Column-wall.—The cuticle on the outside is devoid of adhering foreign matter, except in places where a unicellular green alga is attached, giving rise to the greenish patches seen on the living animal. The sub-cuticla is of regular thickness, but enlarged a little where it communicates with the internal mesoglœa by strands across the ectoderm. The ectoderm is thick ; the nuclei of the individual cells show no regular zonal arrangement ; it is broken up by the strands of mesoglœa into somewhat cubical or spheroidal blocks ; large, thick-walled, highly refractive zooxanthellæ, and occasional large colourless stinging cells are present in places. In the uppermost part of the column the sub-cuticla is absent, and the ectoderm continuous. The internal boundary is often not clearly defined, portions being, as it were, cut off and isolated, and, as still smaller parts, often only individual cells, sunk further into the mesoglœa.

The mesoglœa is very thick, and contains abundant cell-inclosures, and uniformly distributed small cells with granular protoplasm. In the lower parts more especially these take the form of small communicating canals. M⁰Murrich refers to the alteration in histological structure which some of the cells undergo in the mesoglœa, by which they become filled with refractive, deeply staining granules, and suggests (p. 118) that they may be concerned in the formation of the mesoglœa, their granules being particles which will later on be added to the matrix of the mesoglœa. Many of the cells in my sections, generally in limited areas, appear to go a stage beyond, and instead of the granules filling the cells, they become arranged peripherally, giving the appearance of a thickened granular cell-wall, a distinct central nucleus remaining (fig. 6, Pl. XVIII. A).

The endoderm is thin, and contains abundant zooxanthellæ and small stinging cells. A circular endodermal muscle occurs along the greater part of the length of the column. Endodermal bays are met with at different levels, extending nearly as far as the ectoderm, and evidently correspond with the thin annuli noticed amongst the external characters. The endodermal muscle follows the outgrowths for only a short distance. Perhaps the bays serve to give flexibility to the column. The projections seen on the upper part of the column are shown to be due to thickenings of the mesoglœa, and contain a cavity lined with cells continuous with the endoderm, as is recorded by Mᶜ Murrich (p. 192), for the Bahaman forms; further, some polyps do not show any of the ectodermal bays mentioned by Mᶜ Murrich, and by Haddon and Shackleton, while they occur in others.

Sphincter muscle (Pl. XVIII. A, fig. 5).—The single mesoglœal sphincter muscle is strong, elongated, and in longitudinal sections extends nearly across the mesoglœa. Proximally, the mesoglœal cavities are small and circular; distally, they are oval and more elongated; the muscle cells are arranged in different directions, and constitute a very thin layer, the remainder of the cavity being partially occupied with loose rounded cells, or more usually appears as an empty space. For the greater part of its length, the muscle cavities give a vesicular character to the mesoglœa. The appearance is figured by Mᶜ Murrich, but the cavities appear more numerous and not so elongated in the Jamaican form. Some of my preparations show the constriction he refers to, but I have not obtained the long, branching, terminal cavities.

Tentacles.—The tentacles, seen in longitudinal sections, are as long as usual in the Zoanthidæ, and are acuminate in form. The ectoderm is very thick and shows a peripheral zone of colourless, narrow nematocysts and gland cells; below this a zone of deeply staining small nuclei; nearer the mesoglœa abundant pigment granules occur, along with a few scattered nuclei. The mesoglœa is thin, but thrown into fine, long, branching plaits on the ectodermal side for the support of the muscle, to such an extent that, in some sections, the mesoglœa appears as if it had enclosed parts of the ectoderm. This condition is also described for *I. asymmetricus* (1891, p. 685). The endoderm, even in the state of retraction, is much thinner than the ectoderm, an unusual condition in the Zoanthidæ. It contains numerous ordinary zooxanthellæ, and occasionally others with thick, highly refractive walls, such as are found in the ectoderm of the body-wall, and are there also associated with the thin-walled form.

Disk.—The disk much resembles the tentacles, but the ectoderm is not so thick, nor nematocysts so abundant. The nuclei are more uniformly distributed; but the peculiar ectodermal musculature is similar in places. An endodermal muscle also occurs.

Œsophagus.—The œsophagus is considerably folded; the three layers maintain a somewhat uniform thickness, but the mesoglœa follows in places the more deeply folding ectoderm. It is slightly truncated opposite the sulcar directives, the two mesenteries extending from each corner being the only indication of an œsophageal groove. The ectoderm is ciliated; the small, deeply-staining oval nuclei are arranged in a band a little below the surface. Gland cells, and elongated nematocysts, showing a spiral thread, occur sparingly. The mesogloea is thin, and small granular cell-enclosures are scattered throughout.

The endoderm is a low band of cells resembling that of the mesenteries; zooxanthellæ, small nematocysts, and a weak muscle are present.

Mesenteries.—The mesenteries are brachycnemic in type, and the perfect ones are arranged at about equal distances apart all round the œsophagus. Twenty-one pairs are present in one specimen. The endoderm is thin and crowded with zooxanthellæ and small oval nematocysts. A parieto-basilar muscle and a vertically arranged musculature occur on each side. The mesogloea is well developed throughout. Towards the insertion of the mesentery into the body-wall it is thrown into small irregular plaits or pennons; still nearer it narrows a little. A basal canal and numerous irregular vertical canals and cell-enclosures occur the whole length of the mesenteries, continuous in places with those in the mesogloea of the column-wall. The reflected ectoderm rarely occurs, but the mesenterial filaments are met with as usual. Towards the base of the polyp the mesenteries begin to unite with one another, and ultimately form a reticulum-like structure filling the whole of the cœlenteron.

Gonads.—No gonads were present in three examples sectionized.

From the latest researches of Professor M℃Murrich, it appears that the West Indies possess two species of Isaurus, one from Bermuda, identified by him as the *Isaurus tuberculatus*, of Gray (1828), and another, the *Zoanthus tuberculatus*, of Duchassaing (1850), obtained from the Bahamas in the Northrop Collection, and previously collected from Guadaloupe and St. Thomas. In his Bermudan paper (1889 *a*), M℃Murrich, however, considered Gray's form as identical, not only with the Bermudan examples, but also with the *Z. tuberculatus*. Owing to these later results, and the specific name *tuberculatus* being occupied by both forms, he has followed Andres and adopted the term *Duchassaingi* for the Bahaman examples and for those known to Duchassaing and Michelotti.

Professor Haddon and Miss Shackleton (1891) have described as new, a form, *I. asymmetricus*, obtained by the senior author from Torres Straits. In doing this they state (p. 684):—"It is undoubtedly nearly allied to the *Mammillifera tuberculatus* of M℃Murrich. The specific differences are the lesser number and greater size of the tubercles, though their diameter is about the same, and their asymmetrical arrangement; the height of our species is about double that of the West Indian form."

The specimens described above seem to me to unite in a very marked manner the two West Indian and also the Torres Straits examples. I regard the differences in the external appearance of the tubercles, transverse annulations, &c., as largely dependent upon age and method of preservation. Even in the details of microscopic structure, the Jamaican specimens appear to agree very closely, particularly so in the peculiar mesoglœal plaitings of the tentacles and the form of the mesenteries.

Colonies obtained later from Port Antonio convince me that it will be found impossible to maintain the separation of the three species, to such an extent is the form variable in external characters and structure.

Gemmaria, DUCHASSAING and MICHELOTTI, 1860.

Brachycnemic Zoantheæ, with a single mesoglœal sphincter muscle. Solitary, or connected by cœnosarc. The body-wall is incrusted. The ectoderm is usually discontinuous, but may be continuous. Lacunæ and cell-islets are found in the mesoglœa. Diœcious or monœcious.

The only difference between the definition of the genus here given and that in a former publication (1896, p. 142) is in connexion with the gonads. All the species hitherto examined have had the male and female reproductive cells, where present, in different individuals; but in the first representative described below, both ova and spermaria occur on the same mesenteries (Pl. XVIII. A, fig. 8). It has already been shown (1891, p. 623) that a similar monœcious and diœcious condition exists in the genus Zoanthus, and doubtfully in Isaurus.

Gemmaria variabilis, n. sp.

(Pl. XVII. A, fig. 5.)

Form.—Polyps erect, firm, smooth, arising independently from a lamellar cœnenchyme, or from around the base of one another, or may be solitary; often cylindrical in retraction; slightly enlarged and flattened distally, or occasionally narrowing and terminating bluntly; others, mostly long examples, are clavate, being narrow below and expanding above either slowly or more suddenly; transversely wrinkled, especially in spirit specimens. Capitulum with about thirty ridges and furrows. Tentacles acuminate, arranged in two alternating rows of about thirty in each row; the number may vary considerably, forty in each row being counted in one example. Peristome considerably raised; the mouth elongated and slit-like.

In full expansion, the capitulum and disc are much enlarged in proportion to the diameter of the column; and the individuals in a colony are so closely aggregated that, reaching the same level, the margins come in contact, and by mutual pressure produce a polygonal outline, leaving no interstices. Where examples in a colony incrust an irregular surface, or are fixed to the underside of stones, the columns elongate sufficiently to bring all the individuals, with the disc looking upwards, to about the same level. A living colony when fully expanded thus presents the appearance of a mosaic work of brown or green depressed discs, with margins of a dark-brown colour.

When alive, polyps are found under three conditions:—

(1) Retraction, where the disc and tentacles are entirely withdrawn, leaving only a very small central opening.

(2) Partial expansion, with a small portion of the disc visible. This is considerably depressed, and the tips of the tentacles protrude from between the thick capitulum and the disc.

(3) Full expansion, in which the disc is completely exposed and only slightly below the capitular margin, and the tentacles are quite free. In this state the capitula are in contact with one another.

Cœnenchyme present around the base of each polyp, but otherwise not very freely developed, appearing rather as a consequence of the origin of the polyps from one another by basal gemmation, and connecting them only as a flattened band or ribbon; the band may become constricted, and finally the individuals sever their connexion with one another.

Colour.—Lower part of column light buff, upper dark brown. Tentacles usually dark brown, but may be olive or green. Disc in some is dark brown, with green radiating lines, and the peristome a bright green; or the disk may be green and the peristome brown; in others the disc and peristome are both bright green. Œsophagus colourless. The ectoderm containing the brown pigment readily rubs off when handled, the colourless mesogloea, with the enclosed sand grains, being exposed. In alcohol, the brown colour is first extracted, leaving the colony a uniformly dark green; later this gives place to a dirty buff colour.

Dimensions.—The dimensions of the individual polyps vary considerably even in the same colony, being largely dependent upon the position of the polyp in the colony. In the large masses spreading over an even surface, the individuals are all of the same thick-set type and approximately of uniform size. When the colonies are smaller, and the incrusted surface irregular, the specimens in the depressions become elongated in order to attain the same level as the majority. The length of the column of one of the longest is 5 cm., the diameter 1·2 cm.; an average height is 1·5 cm., and diameter 0·7 cm.; diameter of expanded disk 2·3 cm.; tentacles about 0·3 cm. in length. Owing to the rigidity of the column-wall there is not much contraction in preserved specimens.

Locality.—Found growing very abundantly upon rocks and stones in shallow water at Port Henderson, Kingston Harbour. Numerous irregular colonies are to be met with, sometimes one or two feet across; one was over two yards in length, and one to two feet broad. Incrusting sponges grow freely on the cœnenchyme and amongst the polyps, and Ophiuroids meander around.

The specific name has reference to the amount of variation met with in the various external features of the polyps.

Column-wall (Pl. XVIII. A, fig. 7).—The cuticle of the column-wall is thickly coated below with a layer of foreign matter, principally diatoms. The ectoderm is continuous, and presents irregular internal limitations, especially towards the upper part of the column. This is due partly to the presence of incrustations, but also to the ectoderm passing insensibly into the cell-enclosures of the mesoglœa. Numerous zooxanthellæ occur, and occasionally large colourless oval nematocysts, showing the coiled internal thread. The incrustations are sand grains, sponge spicules, and tests of Radiolarians, and extend from the inner border of the ectoderm to beyond the middle of the mesoglœa.

The mesoglœa is thicker above and below than in the middle; numerous cells occur bearing elongated processes, and cell-enclosures of various dimensions uniformly distributed. Fine radiating processes extend from the endodermal boundary, apparently throughout the layer; the large cell-islets contain zooxanthellæ and large oval nematocyts, as in the ectoderm.

The endoderm is of medium height, and contains zooxanthellæ and pigment granules; the circular endodermal muscle is easily distinguished.

Sphincter muscle (Pl. XVIII. A, fig. 7).—The sphincter muscle is single and enclosed in an extended series of small mesoglœal cavities, varying but slightly in size, shape, and distance apart. It is situated nearer the endoderm. The cavities in the upper part are a little larger, and the lining muscle-fibres are arranged in various directions, many being cut obliquely. A few small spherical cells are also present in the middle of the cavities.

Tentacles.—The ectoderm is thick and has an outer layer of small narrow stinging cells, and below this abundant deeply staining oval nuclei and numerous glandular cells. The mesoglœa is broad, and contains isolated cells and foreign incrusting matter. An ectodermal and an endodermal musculature occur.

Disc.—The ectoderm is very broad and contains zooxanthellæ and glandular cells. The mesoglœa is nearly as thick as that of the column-wall, but contains no foreign inclosures; minute cellular strands and a few cell-islets with large oval nematocysts occur in it. In the peripheral part of the disc, the mesoglœa is very thin, while the ectoderm is a little thicker than in the more central region. The endoderm is low and contains zooxanthellæ; a weak endodermal muscle on plaitings of the mesoglœa occurs.

Œsophagus.—The œsophagus is oval-shaped in transverse sections, with a well marked truncated œsophageal groove, the sulcar directives extending from the corners. The ectoderm is thrown into longitudinal folds, not followed by the mesoglœa; twelve occur on each side in one specimen, but there may be as many as 15 or 18. The large colourless nematocysts and pigment granules are present, and a weak nerve layer.

The cells are longer at the groove. The mesoglœa is thin, but thickens towards the same place, and contains cells. The endoderm is low, and shows an outer zone of nuclei and an inner non-staining zone; the endodermal muscle is supported on mesoglœal plaitings.

Mesenteries (Pl. XVIII. A, fig. 9).—The usual brachycnemic condition is present in most; but in two specimens the mesenteries are brachycnemic on one side, and macrocnemic on the other. In most, fifteen perfect mesenteries occur on each side, and the same number of imperfect. In one, twenty-seven pairs in all were present, and in another twenty-eight pairs. Each has an irregularly shaped basal canal a little beyond the origin, and, in the upper part of the column, others extend almost across the mesentery. The basal canal is continued the whole length of the mesentery, and contains zooxanthellæ and large oval nematocysts; it may be divided in the upper part into two or more closely approximated canals.

The parieto-basilar muscles are well developed. Beyond the basal portion the mesenteries are very thin, and the endoderm is crowded with large zooxanthellæ. The imperfect mesenteries are very short proximally, appearing in transverse sections as goblet-shaped projections of the body-wall; the muscle extends all round, while the basal canal is more circular than in the others. The reflected ectoderm and mesenterial filaments are well developed.

Gonads (Pl. XVIII. A, fig. 8).—In one specimen examined, both male and female gonads were found in abundance; sometimes both kinds would occur on one mesentery, while others bore either ova or spermaria. The ova, which evidently were nearly ripe, were scarcely stained with borax carmine, while the spermaria readily took up the pigment.

Cœnenchyme.—In its outer part, the cœnenchyme has numerous inclosures similar to those of the body-wall. Many large cell inclosures and cœlenteric canals are met with, the latter with a very regular epithelial lining and a weak musculature.

Under their genus Gemmaria, MM. Duchassaing and Michelotti describe (1860) four species of Zoanthidæ from the Antilles, viz.:—*G. Rusei*, Duch. and Michel.; *G. clavata*, Duch.; *G. Swiftii*, Duch. and Michel.; and *G. brevis*, Duch. The first has been recovered in the Bermudas by McMurrich (1889), while *G. Swiftii* is shown in the present Paper to belong to the genus Parazoanthus.

I have hesitated considerably as to the identity of the present form with *G. clavata*, but have finally decided that the characters given in the two descriptions of it will not admit of this. The original diagnosis (1850, p. 11) gives the tentacles as about 30, and the later one (1860, p. 331) states the disc and tentacles to be violet.

The Jamaican form also appears to be a larger, more robust species. External characters readily separate it from *G. brevis*. *G. isolata*, described by M⁰Murrich, from the Bahamas (1889), is also evidently quite distinct. It can likewise be distinguished from the other known members of the genus—*G. Macmurrichi*, Hadd. and Shackl.; *G. Mutuki*, Hadd. and Shackl.; and *G. canariensis*, Hadd. and Duerd.—obtained from localities more distant.

Gemmaria fusca, n. sp.

(Pl. XVII. A, fig. 6.)

Form.—Polyps erect, firm, cylindrical, growing in colonies from a thin lamellar cœnenchyme or solitary; smooth above, with sand grains showing through the ectoderm, and scarcely any adhering particles, but many more below. Capitulum with about 30 ridges and furrows, may be slightly more or less; greatly expanded and overhanging in full extension. Tentacles dicyclic, smooth, acuminate, overhanging in extension, short, slightly entacmæous. Outer part of disc overhanging in full extension, giving an umbrella-like appearance, with the radiating mesenterial lines showing through; central portion of disc appears as a rounded elevation with the slit-like mouth at the apex, and is devoid of incrustations. Cœnenchyme spreading and closely incrusting the upper surface of rocks and stones, not very freely developed; exposed surface rough, due to adhering calcareous particles.

The individual polyps in a colony are usually closely apposed at the base, but may be separated a short distance from one another, or may ultimately become isolated. The polyps are practically the same diameter throughout, but may diminish a little below, expanding again towards the base. In retraction the distal part may be slightly swollen and rounded, with a central aperture; the number of capitular ridges, which extend for some distance down the column, is very variable. In preserved specimens the proximal part of the column is slightly wrinkled, but the distal is smooth.

Colour.—Distal part of column, tentacles, and disc dark brown; proximal part of column sand-coloured, often with foreign green matter; œsophagus white.

Dimensions.—Height of column varies from 1 to 3 cm., most are about 2·2 cm.; diameter 1 cm.; inner tentacles 0·15 cm. in length.

Locality.—Colonies and isolated individuals are found growing in considerable abundance attached to coral rock and stones in the very shallow water around

Drunkenman Cay; sometimes the polyps are partially embedded in sand and débris. Numerous young individuals arising directly from the cœnenchyme, or from the base of other polyps, are mingled with the older examples.

Column-wall (Pl. XVIII.A, fig. 10).—The ectoderm is continuous, and not much broken up by incrusting matter. The cuticle is thin, with few adhering foreign bodies. The ectoderm is broad in the distal part of the column, but narrows below, and the nuclei of the cells are uniformly distributed except near the cuticle, a regular columnar epithelium not being formed. Large oval nematocysts occur, and large zooxanthellæ are present in company with small narrow stinging cells and cells containing highly refractive pigment granules. The inner boundaries of the ectoderm are not well defined, and at the capitulum the layer becomes very thick and still more irregular in its internal outline; definite bays or growths into the mesoglœa appear in sections, probably due to the presence of capitular ridges and furrows.

The mesoglœa is of medium thickness, enlarging a little both proximally and distally; the incrustations are limited to the outer portion and the adjacent ectoderm. They occur very sparingly, not interfering with the cutting of thin sections, and consist of calcareous and a few siliceous sand grains, sponge spicules, and an occasional Foraminiferal or Radiolarian test. The mesoglœa contains isolated cells and cell-islets distributed with some uniformity, except in the lower part where an irregular zone of larger inclosures may be found a little nearer the inner boundary. The larger islets contain zooxanthellæ, large oval nematocysts, and occasionally pigment granules similar to those in the ectoderm, from which layer the cell-islets can be seen to originate.

The endoderm is thin, more so than in *G. variabilis*, and contains many zooxanthellæ. The circular endodermal muscle is well developed; fine fibrils from it stretch nearly across the mesoglœa, and others are seen connecting the various cells and cell-islets.

Sphincter muscle (Pl. XVIII. A, fig. 10).—The sphincter muscle is single and mesoglœal. It is long and situated near the endoderm. Proximally it commences in small irregular cavities in groups of two or three, and arranged in a not very regular row. The more distal cavities are much larger, irregular in form, and extend further across the mesoglœa; the muscle fibres are arranged obliquely, and isolated spherical cells occur. The muscle is shorter, the cavities less regular in arrangement, and not in such a single series as in *G. variabilis*; while the upper ones are closer, broader, and more irregular in outline.

Tentacles.—The ectoderm of the tentacles is very thick, and consists of an outer zone of small narrow nematocysts, and an inner zone of zooxanthellæ and nuclei irregularly arranged. A few pigment granules, a number of homogeneous

3 H 2

deeply staining bodies, and glands filled with clear contents, are met with, and an occasional large oval stinging cell. Transverse sections show a well developed ectodermal muscle on mesoglœal plaitings. The mesoglœa has small cells scattered throughout. The endoderm is made up of small regularly arranged cells; an endodermal circular muscle is supported on fine mesoglœal plaitings; and zooxanthellæ are present.

Disc.—The ectoderm of the disc is even thicker than that of the tentacles, and exhibits an outer zone of clear gland spaces and small narrow nematocysts. The deeper part is largely composed of zooxanthellæ, glandular cells or spaces, and an occasional large oval stinging cell. An ectodermal muscle occurs on mesoglœal plaitings, and an endodermal muscle is present. The mesoglœa is thick and contains cells and cell-islets, but is devoid of incrustations.

Œsophagus.—Only a slight œsophageal groove is indicated, the mesoglœa being a little thickened and truncate, and the directives extend from the two corners. Below it is oval-shaped in section, and the ectoderm remains unfolded; distally the latter is thrown into eight or nine well marked folds on each side, only exceptionally followed by the mesoglœa. In a second specimen, the number of folds was fifteen on each side. Immediately on passing, in longitudinal sections, beyond the lips of the mouth, the ectoderm undergoes a great alteration from that of the disc. It is richly ciliated, a narrow zone immediately below is colourless; then follows a thick zone of narrow, closely-arranged nuclei, gland cells, and nematocysts, which together form a dense deeply staining band, extending the whole length of the œsophagus. A zone below this has only a few scattered nuclei, and, in places, the large oval stinging cells and pigment granules. The mesoglœa is a homogeneous layer with rarely an enclosed cell, and the endoderm is extremely thin. A weak endodermal but no ectodermal musculature occurs. Terminally the ectoderm is reflected upwards on the mesenteries for a short distance, and folded in a double pinnate manner, and then descends, constituting the mesenterial filaments.

Mesenteries.—The mesenteries are brachycnemic in arrangement. Twelve perfect pairs occur on each side in one specimen, and sixteen in another. In the middle œsophageal region each is extremely delicate, scarcely showing any enlargement towards the insertion at the body-wall or œsophagus; the usual basal canal is often absent, especially distally. The imperfect mesenteries are broad above, but very short below, not being readily distinguishable in places. The parieto-basilar muscle is present on each side; the retractor muscle is weak. A little beyond the insertion of the mesenteries is the flattened or oval-shaped basal canal filled with deeply-staining tissue, and now and then a large oval nematocyst. The endoderm is poorly developed, and has large zooxanthellæ.

Gonads.—No reproductive cells were present in the examples studied.

The combination of characters in which *Gemmaria fusca* differs from *G. variabilis* are : (1) the uniformly brown colouration, (2) the paucity of the incrustations, (3) the almost absence of basal canals in the upper part of the perfect mesenteries, (4) the appearance of the sphincter muscle, and (5) the generally more delicate structure throughout.

For some time I was inclined to regard these two species as being the same, and it was not until an anatomical examination had been made that their distinction was fully apparent. The sphincter muscle, quantity of incrustations, and other structures, are different. Externally they may be distinguished by their colouration. Though not inclined to regard this character as very constant, amongst numerous colonies I have met with no variations from the type in the present species, nor wholly brown examples of *G. variabilis*. *G. fusca* is longer and more regularly cylindrical than the other, without the same tendency to assume a clavate shape. It is also less rigid, the body-wall not being so thick and incrusted, and young individuals arise more numerously amongst the older polyps. The colonies are smaller and less associated.

Palythoa, LAMOUROUX, 1816.

Brachycnemic Zoantheæ with a single mesogloeal sphincter muscle. The body-wall is incrusted. The ectoderm is continuous. The mesogloea contains numerous lacunæ, and occasionally canals. Diœcious. Polyps immersed in a thick cœnenchyme, which forms a massive expansion.

The above is the definition of the genus given by Haddon and Shackleton (1891, p. 691), who also add a detailed history of its complicated career.

In regard to the specific identification of its members, the genus Palythoa has always been recognized by specialists as one of extreme difficulty and uncertainty on account of its variability in form and the presence of only a few external diagnostic characters. Especially is this the case when, as usually obtains, its representatives are studied as alcoholic specimens in a condition of retraction and shrinkage in variable degrees. Great danger exists under these circumstances in the identification of isolated patches, or of even complete colonies. The external characters one has usually to depend upon are those of the amount of the column of the polyps free from the cœnenchyme, the dimensions, colour, wrinklings, number of capitular ridges and tentacles. All these are, however, very inconstant ; only when a number of examples are obtainable for comparison can much value be placed upon them.

The height of the free portion of the column is mainly dependent upon the extent of retraction of the polyps, and is not a constant specific character. In one portion of a colony the upper surface may be almost uniform, none of the polyps being raised above the general level of the cœnenchyme; in another part the polyps may be quite free for a few millimetres. Again, colonies would be collected which became quite flat on their distal surface, due to excessive retraction as a result of rough handling; but, on coming to observe them later, the polyps had elevated the upper part of their column distinctly above the level of the cœnenchyme, and the whole presented a very different aspect. Such variations have also been noted as occurring on the same colony at one and the same time. The dimensions, especially the height of the column and cœnenchyme, are also very variable, depending largely upon the irregularities of the incrusted rock and the presence of contiguous colonies; one side of a colony may be two or three times the height of another.

The colour is so very similar in all the Jamaican forms I have examined as to be of little assistance. So far I have only met with various tints of yellow and brown; a colony which appears yellowish in the water may become brown on removal. Transverse and longitudinal wrinklings are largely determined by the amount of shrinkage in preservation; alcohol material showing more than formalin. The number of tentacles has been used by Duchassaing and Michelotti as an important aid in distinguishing species. Generally, this can only be of service where one is able to study the forms in the living condition, most colonies appearing to prefer a retracted state after removal from their natural habitat. I have generally found that unless much injured in removal, the polyps open out sufficiently for investigation during the first day in the laboratory, and plunging into formalin has fixed them in this condition. I have studied the capitular or marginal ridges more particularly when the polyps have been thus partially or completely open. In the numerous cases counted, the ridges were always found to correspond with half the total number of the tentacles, *i.e.*, with one cycle.

Although varying somewhat, there is no doubt that the ridges and tentacles are of considerable utility for systematic purposes. How far they may be depended upon will be seen from the following observations made specially upon numerous colonies from the various Cays to test the reliability of the character. Each of the letters indicates a separate colony, or portion of one, and the figures the number of ridges counted upon the individual polyps sufficiently open for the purpose.

Lime Cay :

A.—15, 18, 16, 16, 15, 15, 14, 15, 14, 16, 14, 15, 14, 16, 14, 15, 14, 14, 16.
B.—19, 20, 19, 20, 19.
C.—20, 19, 19, 20, 20.
D.—22, 19, 19.
E.—14, 14, 18, 14, 14, 15, 15.

South Cay :

F.—14, 14, 14, 14, 14, 15, 14, 16, 14, 16.
G.—18, 18, 18, 18, 19, 18, 18, 17, 19, 18, 18.
H.—18, 18, 19, 18, 18.
I.—18, 19, 18, 17, 19.

Drunkenman Cay :

    J.—19, 18, 20, 18, 19, 19, 19, 20, 19, 19, 20, 20.
    K.—17, 17, 19, 18, 18, 17, 18, 17.
    L.—15, 20, 17, 16.

Gun Cay :

    M.—14, 16, 15, 16, 15, 14, 15, 14, 17, 17, 14, 14, 16, 14, 16, 12, 12.
    N.—14, 14, 16, 15, 14, 15, 16.
    O.—16, 16, 17.
    P.—16, 16, 16.
    Q.—17, 16, 15, 14, 16, 14, 18, 16, 16, 14.

Maiden Cay :

    R.—20, 18, 18, 16.
    S.—18, 19.
    T.—16, 16, 16, 17.

From these it will be seen that, in the series represented by B, C, D, G, H, I, J, K, R, S, the numbers vary from about 18 to 20, and in another, represented by A, E, F, M, N, O, P, Q, T, the numbers are roughly from 14 to 17. The average numbers of the capitular ridges of the two groups seem so constant and distinct that I have considered them of sufficient importance to justify a separation into the two accompanying species, and have found at the same time other distinctions which further support the conclusion. Although the genus occurs in such abundance around all the Cays, I have not found characters of sufficient stability to warrant the separation of more than these two species. Other diagnostic features will be discussed in dealing with their anatomy.

Palythoa mammillosa (Ellis and Solander).

(Pl. XVII. A, figs. 7, 8.)

Lapidis Astroitidis sive stellaris primordia,	Sloane, 1707, vol. 1, tab. 21, figs. 1–3.
Alcyonium mammillosum, . .	Ellis and Solander, 1786, p. 179, tab. 1, figs. 4, 5.
Palythoa mammillosa, . . .	Lamouroux, 1816, p. 361, pl. xiii., fig. 2.
Palythoa mammilosa, ...	Milne-Edwards, 1857, p. 304.]
Palythoa ocellata, ...	Duchassaing and Michelotti, 1860, p. 329.
Palythoa mamillosa, ...	Duchassaing and Michelotti, 1866, p. 140, pl. vi., fig. 10.
Palythoa cinerea, ...	Duchassaing and Michelotti, 1866, p. 141, pl. vi., fig. 8.
Polythoa mammillosa, . . .	Andres, 1883, p. 332.
Polythoa (Corticithoa) cinerea, .	Andres, 1883, p. 323.

Form.—Polyps smooth, rigid, cylindrical, arranged in a somewhat rectangular manner; the smooth ectoderm is easily rubbed off exposing the mesogloea below, with a roughened surface due to the foreign incrustations; in retraction rounded or somewhat flattened above, free for a short distance beyond the surface of the coenenchyme.   In the living state, or when preserved in formalin without contraction, the polyps are equally free all round, and so closely arranged that they are separated above only by polygonal dividing lines, none of the coenenchyme being visible (fig. 7).   Specimens which have been preserved in alcohol and in which shrinkage has taken place are not equally free on all sides, but connected with one another by four (may be three or five when the polyps are not arranged in a rectangular manner) higher, occasionally grooved, ridges of coenenchyme, and rounded depressions of coenenchyme, are seen in the spaces between (fig. 8).   A central, slightly depressed aperture remains in retracted polyps, and occasionally three to six longitudinal wrinklings along the free portion of the wall of the peripheral polyps are present in specimens preserved in alcohol, and also transverse wrinklings.   The amount of the free portion varies according to the state of extension or retraction of the polyps in a colony.   Usually in complete retraction about 0·4 cm. are free; in partial retraction, when the full capitular ridges can be counted, and in full expansion, about 0·6 cm. are free. In almost complete retraction the capitular ridges are wedge-shaped with very narrow furrows; as the polyps slowly open, the ridges become more convoluted or laterally undulating, and finally appear as so many acute marginal denticulations. These, as already shown above, are usually from 18 to 20 in number.   The polyps of three other colonies from South Cay had a very regular number of ridges as follows :—

A.—19, 18, 18, 18, 18, 18, 18, 19, 18.
B.—19, 18, 18, 18, 18, 18, 18, 18, 19, 18, 19, 18, 19.
C.—18, 21, 18, 19, 18, 18, 18.

Tentacles very short, smooth, acuminate, dicyclic, inner row opposite the marginal denticulations, slightly entacmæous, overhanging in extension, 18 to 20 in each row.   Disc cup-shaped in partial, and saucer-shaped in full, extension, but with the central part appearing as a dome and bearing the slit-like mouth at the apex.   The peripheral zone of the disc is thin-walled, pellucid, smooth, devoid of incrustations, and raised into elevations and grooves corresponding with the number of tentacles, of which it appears as a continuation.   In full extension it is flat or may be arched over; in partial extension it is nearly vertical.   The central part of the disc is smooth, but contains a few incrustations.   The species usually occurs in small, rather high colonies, closely associated with one another, but separated by deep channels.   The incrusting base is much smaller in area than the distal surface, the peripheral polyps being arranged obliquely or radiately. New individuals arise mostly along the side of those forming the periphery of a colony.

Colour.—Cœnenchyme and column pale yellow or cream colour, sometimes brown; tentacles and furrowed portion of disc pellucid; middle of disc grey, due to presence of sand-grains; lips white.

Dimensions.—Average height of polyps 1·3 cm., may be only 0·6 cm. or 1·8 cm.; diameter of flat expanded disc 1·2 cm.; average diameter of columns 0·6 cm.; distance of centres of contiguous polyps in retraction about 0·7 cm.; inner tentacles 0·2 cm. long. Colonies of various sizes are met with, but usually from 3 to 8 cm. across.

Locality.—Jamaica: found in abundance firmly encrusting the coral-rock in shallow water, mostly in the region of the breakers, at the Cays outside Kingston Harbour, and at other points around the coast.

Range.—Guadaloupe and St. Thomas (Duchassaing and Michelotti).

Column-wall (Pl. XIX., fig. 1).—The column-wall of the individual polyps is separable from the cœnenchyme for only a short distance at the upper surface of the colony. The two are not very thick compared with the size of the polyps. In transverse sections the cœnenchyme may be from 0·1 to 0·2 cm. in thickness, while the polyps are from 0·6 cm. to 0·3 cm. in diameter.

The ectoderm is a thick layer, and continuous over the outer surface of a colony; a cuticle, devoid of foreign adhering matter, is present. Its internal limitations are occasionally irregular, due to the incrusting sand-grains; these latter are limited to the inner part of the ectoderm and to the mesoglœa. The outer portion of the ectodermal cells is largely glandular; the middle and inner parts contain the nuclei, numerous zooxanthellæ, and occasional large oval colourless nematocysts showing the internal thread.

The mesoglœa varies in thickness, appearing in sections as a matrix in which the cylindrical polyps are embedded. The incrusting material is practically distributed throughout; in the periphery of a colony however it is rather limited to the outer half. The foreign objects are mostly calcareous, but a few siliceous sponge spicules and an occasional Foraminiferal and Radiolarian test may be present. Abundant small and large cell-islets or short canals are distributed with considerable uniformity; the latter in some sections exhibit somewhat of a concentric arrangement around the individual polyps, and may be seen communicating with the canals in the mesenteries. In others, a canal appears opposite nearly all the mesenteries. The islets contain zooxanthellæ and large oval nematocysts; the smaller islets have the protoplasm exhibiting peripherally a fine morula-like appearance, with a central more deeply staining nucleus. A similar condition is described under *Isaurus tuberculatus* (p. 347). Fine threads connect one group of cells with another.

The endoderm is a uniformly thin layer, and contains zooxanthellæ. A weak circular endodermal muscle is present.

Sphincter muscle (Pl. XIX., fig. 1).—The sphincter muscle is single, mesoglœal, and contained in an elongated series of irregular cavities, situated near the endoderm; small cavities occur along with larger ones, all forming an irregular row.

Tentacles.—The ectoderm is very thick, consisting of an outer zone of small, elongated, narrow nematocysts, and an inner one with deeply-staining nuclei, zooxanthellæ, and pigment granules. The ectodermal muscle is well developed on mesoglœal plaitings. The mesoglœa in places is rather thick, and contains cell-islets. The endoderm is somewhat high, nearly filling up the lumen; the circular muscle is readily seen, forming a very regular layer in longitudinal sections.

Disc (Pl. XIX., fig. 4).—The peripheral grooved portion of the disc has a very thick, highly glandular, sinuous ectoderm, containing zooxanthellæ, small peripheral nematocysts, and occasional deeper oval ones; also a well developed radial muscle. The mesoglœa follows the projecting fold, and becomes very thick, and may contain a few sand-grains; between the folds it is very thin. The endoderm is the same as elsewhere. In longitudinal sections of contracted specimens the part next the tentacles is thick; it then becomes delicate, and again enlarges towards the lips where incrustations occur in the mesoglœa. An endodermal muscle is present.

Œsophagus (Pl. XIX., figs. 2, 3).—The œsophagus in different sections is either an elongate or a shortened pear-shape, with a well defined œsophageal groove. The ectoderm is richly ciliated; a zone of closely-arranged narrow nematocysts occurs peripherally, while the nuclei are arranged mostly in a middle zone; pigment matter occurs in the deeper parts, abundantly in some, sparingly in others. The ectoderm in most is thrown into folds which vary in number, but are generally between eight and eleven; in some sections the ectoderm is unfolded.

The mesoglœa is thin, becoming a little thicker at the groove; it does not follow the foldings of the ectoderm.

The endoderm is similar to that in the column-wall, but is slightly deeper between the mesenteries. It differs from that of the mesenteries in having little or no pigment matter.

Mesenteries (Pl. XIX., figs. 2, 3).—The mesenteries in most cases present the usual brachycnemic type, but irregularities may occur, and opposite sides have not always the same number of pairs. The usual arrangement is that of ten perfect mesenteries on each side, but in one polyp there are ten on one side and nine on the other; in another seven and nine, arranged as shown in fig. 3. The fundamental distinction of the Zoanthidæ into Brachycneminæ and Macrocneminæ is departed from in the sections of two polyps represented. In fig. 2 it is seen that the normal brachycnemic arrangement holds on the left side, while the macrocnemic is present on the right side. This is also the case in fig. 3, only the order is reversed.

The manner of the connexion of the mesenteries to the œsophagus is best shown in fig. 2. Beyond the sulcar directives there is a considerable interspace before the other mesenteries are reached, and then the interspaces are about equal. The mesenteries are very thin near their attachment to the column-wall, but enlarge a little to form the basal canal. The imperfect mesenteries do not project far. The endoderm contains zooxanthellæ and pigment matter, and the mesoglœa is extremely thin. The parieto-basilar muscles are well developed. The basal canal is usually rounded, and contains numerous large oval nematocysts. The reflected ectoderm and mesenterial filaments present the usual structure. The digestive endoderm is very thick and granular.

Gonads.—No gonads were present in numerous examples sectionized.

Cœnenchyme.—The basal portion of the cœnenchyme is very crowded with canals in communication with the basal canals in the mesenteries and containing pigment granules and large oval nematocysts. Cœlenteric canals connect one polyp with another.

This species, first described by Ellis and Solander, is one of the two original representatives of the genus *Polythoa* of Lamouroux. The material upon which it was founded was originally obtained by Sir Hans Sloane from Jamaican waters, probably about the year 1687, when Sloane visited the island. The specimens were deposited by him in the British Museum; the collections of the famous naturalist and physician forming the nuclei of that national institution. Sloane, however, in his "Voyage," which deals largely with the Natural History of Jamaica, has no description of the objects beyond that given on the plate containing his three figures, "*Lapidis astroitidis sive stellaris primordia,*" implying that this, along with the *Alcyonium ocellatum,* of Ellis and Solander, are the beginnings of the stony star-like corals, so abundant in the seas around.

Ellis and Solander first named, described, and again figured Sloane's specimen. Although their description, "This whitish leather-like Alcyonium is spread over rocks, with many convex teat-like figures, hollow in the middle, with a faint star-like appearance, and united close together," is rather incomplete for purposes of identification, still the excellent figure they give of a colony leaves me little hesitation in considering the form described above as the same these two authors had under consideration. The dimensions, amount of the polyps not immersed in cœnenchyme, and the general form of the colony well agree. I feel all the more assurance in this seeing that similar specimens may be collected in abundance from what we may regard as the original locality of the type. It is not at all improbable that Sloane obtained his examples from precisely the same Cays, these being, as already mentioned, the usual and most favourable spots for marine collectors.

Duchassaing and Michelotti (1860) describe as *P. ocellata* a form which, in their later paper (1866), they place under *P. mamillosa.* They also regard the *Corticifera flava* of Lesueur as a variety. It seems pretty evident that these authors, taking the number of tentacles as a criterion, introduced some little confusion, so that it is now very difficult, if not impossible, to ascertain what forms they are really describing.

There is nothing appearing in the original description and figure of *P. cinerea* which is not met with in the large amount of material of *P. mammillosa* which has come under my observation, the colour, form of the original polyps, and incrustations of the latter presenting all the variations ascribed to the former, while the number of tentacles exactly corresponds.

The species is readily distinguished *in situ* from the next one by its habit of growth, occurring mostly in numerous, closely associated, irregularly shaped, small, high colonies, separated by channels 2 or 3 cm. across. The colonies are usually from 8 to 10 cm. in diameter, but may be more. The individual polyps are also larger, and appear to open more readily and constantly, and to have a greater free distal portion.

The larger number of capitular ridges, tentacles, and corresponding mesenteries is evidently constant. The variations in transverse dimensions are more clearly indicated in sections. In the present species the diameter is often 0·6 cm., while in the next it is rarely more than 0·35 cm.

Histologically I have not been able to detect much specific difference. Numbers of sections have been examined from various colonies, some with the incrustations *in situ* and others decalcified. Although the incrustations are abundant and uniformly distributed throughout the colony, the mesoglœa is apparently not so crowded with them as in *P. caribæa.*

The basal canals appear more rounded, and perhaps the internal pigmentation is not so dense in the present example; also, as shown in the figure, the cavities of the sphincter muscle are not in such a regular row.

Palythoa caribæa, DUCHASSAING and MICHELOTTI.

(Pl. XVII. A, fig. 9.)

Palythoa caribæorum,   Duchassaing and Michelotti, 1860, p. 329.
Palythoa caribæa, .   Duchassaing and Michelotti, 1866, p. 141, Pl. vi.,
                                fig. 11.
Polythoa (Monothoa) caraibeorum, Andres, 1883, p. 322.

Form.—Polyps smooth, rigid, cylindrical, closely associated and arranged in
an irregular manner, usually free from the cœnenchyme for a short distance,
free portion rounded or conical in retraction ; in very strong retraction, the
upper surface of the colony may be nearly flat; no wrinklings in specimens
preserved in formalin. Capitular ridges and furrows variable, usually from 14 to
17. The following numbers counted on two colonies will indicate the amount of
this variation :—

> A.—15, 14, 16, 15, 14, 16, 14, 16, 15, 15, 15, 15, 14, 15, 15, 15, 15, 15, 15, 16, 15, 14,
>   16, 16, 17, 16, 17, 15, 14, 17, 15, 15.

> B.—14, 18, 16, 14, 17, 15, 17, 17, 15, 17, 14, 15, 15, 14, 15, 14, 16, 15, 16, 16, 16, 16,
>   16, 17, 17.

Tentacles dicyclic, smooth, pellucid, very short, acuminate, slightly entacmæous,
inner row opposite capitular ridges, overhanging in full extension, vary from 28
to 34 in number. Disc considerably depressed in partial extension, overhanging
in full extension, cup-shaped or saucer-shaped, according to amount of extension ;
divisible into two portions: an outer, thin, transparent, non-incrusted, circular
area with rounded ridges and furrows corresponding to the number of tentacles,
and a dome-shaped central portion, with a few minute incrustations, and bearing
the slit-like mouth at the apex.

The polyps are arranged very closely, and the amount of cœnenchyme connecting the individuals is thin. At the periphery of the colonies, the outlines of
the different marginal polyps are clearly indicated. New individuals appear to
arise between previously existing ones. The colonies are usually very extensive,
irregular in outline, and often incrust very uneven surfaces, the height of the
polyps varying accordingly, so as to produce a regular undulating surface above.

Colour.—In the living condition, a pale yellow or cream colour, or sometimes
brown, white when the ectoderm is rubbed off, lips white. In specimens preserved
in formalin a curious change is effected. Nearly the whole of the upper surface
of the colony may become a brick-red colour. The capitular ridges, however, for

some distance down the column, are quite colourless and hence readily counted. The tentacles, inturned disc, and edges of the mesenteries are likewise altered in colour.

Dimensions.—The length of the polyps and the thickness of the cœnenchyme differ very much, may vary from 0·3 or 0·4 cm. to 1·8 cm., usually about 0·7 cm.; diameter of disc in partial extension 0·5 cm., in full extension 0·9 cm., in retraction 0·4 cm.; distance of centres of contiguous polyps 0·5 cm.; height of free portion above the level of the cœnenchyme in partial retraction about 0·5 cm.; tentacles about 0·2 cm. long; diameter of polyps in section 0·35 cm.

Activities.—Quantities of bubbles of gas are given off when the colonies are exposed in the water to the direct rays of the sun. The polyps do not appear to open so readily as in *P. mammillosa.*

Locality.—Jamaica: Numerous colonies form flat expansions covering considerable areas of coral rocks, at all the Cays outside Kingston Harbour.

Range.—St. Thomas (Duchassaing and Michelotti).

Column-wall (Pl. XIX., fig. 5).—The lower boundary of the column-wall of the individual polyps in a colony can not be distinguished from the cœnenchyme in which the polyps present the appearance of being embedded, but above it is quite distinct. The ectoderm is continuous, and spreads as a uniform layer over the surface of the whole colony; a thin, well defined cuticle occurs on the outside. It is not readily separable from the mesoglœa, appearing to pass insensibly into the cell-enclosures of the latter; narrow elongated nematocysts occur, as well as very large oval nematocysts, which do not stain; zooxanthellæ are present; foreign incrusting material is met with only in the deeper part of the ectoderm.

The mesoglœa is densely crowded throughout its whole thickness with calcareous sand-grains; a few siliceous sponge spicules, Radiolarians, and rarely a Foraminifera occur; most of the material can be dissolved out by acids. Isolated cells and large and small cell-islets are scattered throughout; the large nematocysts, pigment granules, and densely staining tissue fill up the islets.

The endoderm is very thin and uniform in height, except in the upper region where the mesenteries are closer, when the endoderm elongates in the middle and appears triangular in section. It contains abundant granular pigment matter and zooxanthellæ; a weak endodermal muscle is present on slight plaitings of the mesoglœa, especially in the upper region.

Sphincter muscle (Pl. XIX., fig. 5).—The single sphincter muscle is contained in a very regular series of small mesoglœal cavities; proximally they are situated close to the endodermal border, but are more central above, where also the cavities are not in such a single series and become a little larger. The muscular lining is thick, but does not quite fill the cavities.

Tentacles.—The tentacles have a very broad ciliated ectoderm crowded with narrow elongated nematocysts, zooxanthellæ, and pigment granules ; the mesoglœa and the endoderm are thin. The longitudinal ectodermal muscle is well developed on small mesoglœal plaitings.

Disc.—The ectoderm of the disc is broad and contains nematocysts, zooxanthellæ, and pigment granules in the deeper parts. The mesoglœa thickens towards the middle, and incrustations are there present. The endoderm is like that of the column-wall.

Œsophagus.—The outline of the œsophagus varies in different regions and in different specimens. In most polyps it is the usual pyriform, truncated shape, with the ectoderm thrown into seven or eight longitudinal folds on each side, and the œsophageal groove well marked and occupying nearly one-third of the transverse diameter ; but in others, it may be almost circular in outline with none of the folds showing. The ectoderm is very thick, stains deeply, is richly ciliated, and loaded internally with yellow pigment granules, and outwardly with elongated nematocysts. The mesoglœa is narrow, thickening a little at the groove ; the endoderm is like that of the column-wall.

Mesenteries (Pl. xix., fig. 7).—The mesenteries present the usual brachycnemic arrangement in most cases ; but, as already mentioned, irregularities may occur, so that a polyp may be brachycnemic on one side and macrocnemic on the other, while one polyp has been met with which has the latter arrangement on both sides. The number of pairs is variable, and the two lateral halves do not always correspond. In a portion of one colony two polyps have eight perfect mesenteries on each side ; two have eight on one side and seven on the other ; while another has six on one side and eleven on the other. The imperfect mesenteries are well developed. The endoderm is very thick, and contains zooxanthellæ, nematocysts, and abundant pigment matter. The parieto-basilar muscle is clearly seen on each side, but the retractor muscle layer is scarcely distinguishable. The mesoglœa is extremely thin, except towards the column-wall, where the canals or sinuses extend nearly the whole vertical length ; they occupy almost the whole transverse width in the uppermost region, but are elongated or oval below. The basal canals are well developed in both the perfect and imperfect mesenteries, and are crowded with oval nematocysts, and pigment particles, and connected below with the sinuses in the cœnenchyme. The ectoderm of the œsophagus is reflected and folded on the mesenteries. The endoderm on the lower part of the mesenteries is enormously thickened and loaded with granules ; the mesenterial filaments become nearly circular.

Gonads (Pl. xix., fig. 6).—Spermaria, arranged in vertical and transverse rows, were present in the mesenteries of some of the polyps examined.

I identify the abundant Jamaican material with Duchassaing and Michelotti's species, mainly from the number of tentacles which these authors give, viz., thirty to thirty-two; these coming nearest to those indicated above. The figure which they give is of a dried specimen with all the polyps withdrawn to their full extent.

At first I considered it to be *Alcyonium ocellatum*, Ell. and Sol., obtained along with *A. mammillosum* by Sloane from Jamaica, there being nothing in the original descriptions and figures which is not met with in the specimens I have examined. M⁰Murrich has, however (1889, p. 120), appropriated this name for some small colonies from Shelley Bay, Bermudas. He does this upon very slender grounds, this being the name given it by the collector. He has very kindly compared the Jamaican examples with those from the Bermudas, and states that they are quite different, especially in their anatomical characters, although acknowledging that it would seem as if they were the true *P. ocellata*. Under the circumstances, however, it seems best that M⁰Murrich's identification should stand, and to allocate Duchassaing and Michelotti's name, with which the material very closely agrees. Andres (1883, p. 323) is evidently acting contrary to these two authors in considering the *Hughæa caraibeorum* of Duchassaing as a synonym of this species, as, in the "Mémoire" (p. 315), they place it in the genus Paractis.

In a quantity of colonies, it can easily be separated from *P. mammillosa*, not only by the average number of capitular ridges and tentacles, but by the differences in size of the polyps, those of the present species being smaller and more closely aggregated than the former. Usually the colonies are flatter, and cover larger areas. The polyps generally retract to a greater degree, so that the upper surface of the colony becomes more uniform.

Sub-family.—MACROCNEMINÆ.

Epizoanthus, Gray, 1867.

Macrocnemic Zoantheæ, with a single mesoglœal sphincter muscle. The body-wall is incrusted. The ectoderm is usually continuous, but may be discontinuous; cell-islets in the mesoglœa. Diœcious. Polyps connected by cœnenchyme, which may be band-like, incrusting, or greatly reduced, as in the free form.

The genus *Epizoanthus* is defined as above by Haddon and Shackleton (1891, p. 632) accompanied by a full account of its history. They recognize twelve species from various parts of the world, and four doubtful forms. Some of the representatives of the genus form incrustations over the surface of univalve shells inhabited by hermit-crabs, the shells being ultimately dissolved away. The colony, known as a carcinæcium, retains somewhat the form of the shell, and contains the crustacean still within.

Epizoanthus minutus, n. sp.

(Pl. xvii. A, fig. 10.)

Form.—Polyps cylindrical, rising obliquely or vertically from a thin, incrusting, ribbon-like cœnenchyme. In complete retraction rounded above with a small aperture remaining, but no capitular ridges and grooves distinguishable. Slightly enlarged towards the base; about the same height as breadth in retraction, not much more in extension; surface rough, covered with very fine sand grains; occasionally with slight transverse wrinklings. In partial retraction swollen and flat above, with the wedge-shaped, acute, capitular ridges and furrows visible, and the slit-like mouth showing. In full extension the upper part of the column is spread out and the middle constricted; margin of column with fifteen or sixteen or twenty-one denticulations, each with parallel sides and a straight free edge, giving a castellated appearance. Disc much depressed, cup-shaped, transparent, with lines of attachment of mesenteries showing through; mouth elevated. The disc, as usual, is divisible into a grooved outer part forming the walls of the cup in extension, and appearing as a continuation of the united bases of the tentacles, and an inner, smooth, flat or slightly elevated, central part bearing the mouth in the middle. Tentacles dicyclic, thin, transparent, elongated, slightly swollen and rounded at the tips, outer series alternating with the denticulations, entacmæous, overhanging in extension, generally thirty or thirty-two in number, but occasionally forty-two. Cœnenchyme thin, incrusting, ribbon-shaped or irregularly expanded where the polyps are closer; surface same as that of column-wall. Polyps arise independently, and may be considerably separated or more closely grouped.

Colour.—Column-wall and cœnenchyme are a dirty brown, the colour of the sand particles; denticulations with white margins; disc brown, with darker radiating lines; tentacles transparent, several series of dark patches are present, more pronounced on the outer row, tips almost opaque white.

Dimensions.—Height of polyps in extension 0·6 cm., in contraction 0·35 to 0·2 cm. Diameter in extension 0·3 cm., in retraction 0·25 to 0·2 cm. Length of tentacles in full extension 0·4 cm.

Locality.—Found growing in abundance near the margin of one of the valves of a living Pinna shell, collected towards the eastern extremity of Kingston

Harbour in water of not more than half a fathom in depth, and only a few yards from the shore. The polyps are very sensitive and active, retracting immediately on being touched.

Column-wall (Pl. xx., figs. 1, 2).—The outline of the column-wall, owing to the presence of incrusting material, is very irregular in sections, especially in the lower part; in the region of the marginal denticulations it is sinuous, and thicker, and the incrusting matter is aggregated opposite the inner circle of tentacles. Where perfect, the ectoderm is continuous; in most places, it is broken up or absent. It is covered on the outside by a cuticle with an adhering layer of foreign material, mostly diatom frustrules and fine mud.

The mesoglœa varies in thickness, being much better developed proximally. It contains isolated cells with long processes, cell-islets, and irregular communicating canals. The incrustations are sparsely distributed, and are mostly siliceous sand grains and a few sponge spicules.

The endoderm is very thin and regular, and the transverse muscle is well developed.

Sphincter muscle (Pl. xx., fig. 1).—The single mesoglœal sphincter muscle is small, and formed in a few, irregular, closely set cavities, extending about half way across the mesoglœa, and situated just at the base of the outer row of tentacles. The lining muscle-fibres are weak, and other rounded cells partially fill up the cavities.

Tentacles (Pl. xx., figs. 2, 3).—The ectoderm of the tentacles is thick compared with the two other layers, and the ectodermal muscle is well developed on small mesoglœal plaitings ; numerous small oval nematocysts occur, and pigment granules in places. The nervous layer is clearly distinguished between the nucleated zone and the muscle fibres, and sends connecting strands to each. The mesoglœa and endoderm are both very thin. An endodermal muscle layer is present, seen in longitudinal sections.

Disc (Pl. xx., fig. 2).—The structure of the disc is much like that of the tentacles, but the ectoderm is not so well developed, and its outer grooved portion is in places loaded with pigment granules.

Œsophagus.—In extended specimens the œsophagus is almost circular in outline ; the œsophageal groove is quite shallow. In longitudinal sections the wall is thrown into transverse folds. The ectoderm is a very regular, ciliated layer, with abundant gland-cells and a few small nematocysts; pigment granules occur in the deeper parts ; it is reflected above the lower termination of the œsophagus, and below forms the mesenterial filaments in the usual manner. The mesoglœa and endoderm are very thin, especially the latter.

Mesenteries (Pl. xx., fig. 4).—Sixteen pairs of mesenteries, presenting the usual macrocnemic arrangement, occur in one specimen, and are all very thin except near their origin and where fertile; the imperfect are very short; the parieto-basilar muscle is developed along each side; no basal canals, or only traces of them, are present. The mesenterial muscles are seen on slight plaitings; pigment granules occur in groups on the endoderm. The digestive endoderm is thick, and large oval nematocysts are embedded in it, along with groups of pigment granules. The imperfect mesenteries have the muscle fibres extending all round. In the distal region, just below the œsophagus, the mesoglœa at the origin of the mesenteries is rectangular, but proximally it becomes goblet-shaped, the part produced beyond in perfect mesenteries being extremely thin. Proximally the mesenteries are branched.

Gonads (Pl. xx., fig. 4).—Spermaria apparently enclosed in endoderm were met with in abundance in two specimens.

This species is most closely allied in outward appearance to the well known European *Epizoanthus Couchii* (Johnston), Hadd. and Shackl. Obvious differences occur in the number and form of the capitular denticulations, the Antillean representative having fifteen, sixteen, or twenty-one, truncated at their free edge; while the older species has twelve or fourteen triangular teeth. The tentacles differ in a corresponding manner. Histological characters indicate further distinctions. It is readily separated from the seven other species examined by the two authors mentioned above, and also from the "Challenger" species. Of American forms it appears to bear a close relation to the incompletely described *Epizoanthus Americanus*, Verr. (1869), from Panama.

Parazoanthus, Haddon and Shackleton, 1891.

Macrocnemic Zoantheæ, with a diffuse endodermal sphincter muscle. The body-wall is incrusted. The ectoderm is continuous. Encircling sinus as well as ectodermal canals, lacunæ, and cell-islets in the mesoglœa. Diœcious. Polyps connected by thin cœnenchyme.

This genus, with the above definition, was established by Haddon and Shackleton (1891, p. 653), to include macrocnemic Zoantheæ with a diffuse endodermal muscle, forms which previously had been referred by Hertwig and Erdmann (1888, p. 35) to the genus *Palythoa*. The authors recognize five species examined by them, and two described by Hertwig and Erdmann. The combination of anatomical characters renders it a well-defined genus.

Carlgren (1895) has shown that the genus *Gerardia*, Lac.-Duth., formerly included, with some hesitation, amongst the Antipatharia, is closely allied to the present genus, differing only in the presence of a strongly developed horny skeleton.

3 K 2

Parazoanthus Swiftii (Duchassaing and Michelotti).

(Pl. xvii. A, fig. 11.)

Gemmaria Swiftii, . . . Duchassaing and Michelotti, 1860, p. 331, pl. viii., figs. 17, 18 : 1866, p. 138.

Polythoa (str. s.) axinellæ, Andres, 1883, p. 311, pl. x., fig. 7.

Form.—Polyps very short, cylindrical, mammiform in retraction, erect, firm, smooth, rising from small band-like branching patches of cœnenchyme incrusting the surface of a sponge ; sometimes the polyps are arranged in a single linear series, at other times the cœnenchyme is expanded, and two or three individuals may occur abreast. Capitulum with twelve serrations at the apex. In partial retraction these appear as so many wedge-shaped ridges, with intervening furrows, around a central orifice ; in full retraction the capitular ridges are scarcely visible, and the polyps are rounded above.

Tentacles minute, entacmæous, acuminate, dicyclic, twelve in each row, the outer alternating with the serrations, overhanging in extension. Disc concave, transparent, with mesenteries showing through ; mouth slit-like and capable of considerable eversion ; lips crenate ; œsophagus shows longitudinal mesenterial lines ; oral cone may be considerably elevated. The usual condition of the polyps appears to be that of retraction.

The cœnenchyme is smooth, thin, in irregularly shaped meandering ribbons or patches firmly incrusting and partially embedded in the sponge.

Colour.—Cœnenchyme and column-wall are a bright orange yellow, lighter on the upper part of the column ; tentacles are pale yellow ; disc a darker, and lips a bright yellow. The parts are sand-coloured where the ectoderm is rubbed off. The bright orange colour gives to the colonies a marked contrast with the dark green sponge.

Dimensions.—Height of polyps above the cœnenchyme varies from 0·15 cm. to 0·3 cm. ; diameter of expanded disc 0·4 cm. ; diameter of column in contraction 0·2 cm.

Locality.—Jamaica : Obtained growing in small colonies on a large, erect, blackish-green, branching sponge collected in water of about two fathoms around Rackum Cay ; also from the shallow waters S. W. of Lime Cay, living on the same species of sponge.

Range.—St. Thomas (Duchassaing and Michelotti).

Column-wall (Pl. xx., fig. 5).—The column-wall is very thick. The cuticle is well defined; the ectoderm continuous and variable in height, with irregular internal limitations. Excepting a narrow zone immediately below the cuticle, the ectoderm cells are crowded with abundant yellow pigment granules of various sizes and numerous medium-sized, oval, colourless nematocysts. It is practically free from inclosures, these being limited to the mesogloea.

The mesogloea shows a very marked division into two parts. The outer is a thick layer of variable dimensions, and crowded with foreign inclosures and abundant yellow pigment granules, limited internally by the encircling sinus. This latter is broken here and there by strands of mesogloea, and has very irregular limitations; the cavities are filled with deeply staining tissue and pigment. The inner layer of the mesogloea is clear and nearly homogeneous, devoid of incrustations and pigment granules, and plaited internally to support the endodermal muscle. The incrusted part in retracted specimens is enlarged a little below the middle of the column, and contains cell-enclosures. The incrustations consist of siliceous and a few calcareous sand grains, and sponge spicules. The mesogloea is too crowded with incrustations and pigment matter to allow of any connecting canals which may be present between the ectoderm and the encircling sinus being distinguished.

The endoderm cells are high, especially between the mesenteries, and contain abundant yellow pigment spheres and granules; a little below the middle of the column, they give rise to a well developed circular endodermal muscle supported on folds of the mesogloea.

Sphincter muscle (Pl. xx., fig. 5).—The sphincter muscle is diffuse and endodermal, and formed as a greater concentration of the ordinary endodermal muscle of the column-wall. Distally it is so deeply sunk in the folds of the mesogloea that in some sections it appears to be entirely cut off from the endoderm, and to become a mesogloeal muscle enclosed in separate cavities.

Tentacles.—In transverse sections of retracted polyps, the tentacles are so closely arranged as to become polygonal in outline. The ectoderm is thick, and has an outer zone of narrow nematocysts capable of staining; below is an irregular zone of pigment granules. An ectodermal muscle on slight mesogloeal plaitings is seen in transverse sections. The mesogloea is only a thin layer, and internally is thrown into folds for the support of the circular endodermal muscle.

The endoderm has abundant pigment spheres and fills the lumen in contraction. The distinction between the granular pigment matter in the ectoderm and the spherical form in the endoderm, although of the same colour, is very marked. A similar difference is found in the ectoderm and endoderm of the column, but not to such a degree as in the tentacles. All the three layers of the disc are but little developed, presenting a marked contrast to the tentacles.

Œsophagus.—The ectoderm of the œsophagus is richly ciliated and folded; the deeply staining nuclei are arranged in a middle zone; narrow nematocysts occur and large deeply staining granular gland-cells. The mesoglœa is very narrow. The endoderm is thick and crowded with pigment spheres. A sulcar groove occurs, and here the mesoglœa is much thickened, but contains no cell-enclosures.

Mesenterics (Pl. xx., fig. 6).—Twelve pairs of mesenteries, macrocnemic in their arrangement, are present. In the upper region each is a little narrow at its insertion in the column-wall, but the mesoglœa thickens rapidly ; only for a short distance in the œsophageal region, but more below. In the perfect mesenteries, the mesoglœa beyond is very thin, and appears to alter in character so that it takes the stain better. There are no basal canals nor any cell-enclosures in the upper region ; but lower two or three short canals, or there may be only cell-enclosures with pigment granules, appearing in the thickened part of the mesenteries. The endoderm is like that of the column-wall. Below the œsophagus it thickens enormously, and contains much pigment and granular matter ; the mesenterial filaments are well developed and branched. In these, the zone of nuclei stains deeply, and occasional very deeply staining glandular cells are present along with nematocysts and much pigment matter. The parieto-basilar muscle is well marked on each side of the mesentery, extending a very little along the column-wall. In the imperfect mesenteries, the musculature extends the whole way round ; in the perfect mesenteries, scarcely any distinction can be made in the musculature of each side, and the mesoglœa is finely plaited.

Gonads (Pl. xx., fig. 6).—All the specimens examined from one colony contained abundant ova, present only on the perfect mesenteries, and associated with much pigment matter and enormously thickened endoderm.

Cœnenchyme and *Base* (Pl. xx., fig. 5).—The proximal surface of the base and cœnenchyme, in contact throughout with the sponge, has a thin ectoderm crowded with yellow pigment granules. The ectoderm of the upper surface of the cœnenchyme is thick, and resembles that of the column-wall.

The mesoglœa is rather thick, and its inclosures are similar to those of the column-wall, but with a larger proportion of sponge spicules ; cell-inclosures are numerous, and contain pigment granules. The endoderm of the base of the polyp is very thin, and contains pigment spheres and granules.

This species was first described by Duchassaing and Michelotti from specimens obtained at St. Thomas. Of their figures (references to which are omitted from the " Explication des Planches "), fig. 18 gives an approximate representation of the appearance of the colonies on the sponge; but fig. 17 is probably erroneous in the number and appearance of the capitular ridges and furrows indicated. Eight of these are shown, while in every case I have found twelve. In their later paper (1866, p. 138) they state the number of tentacles to be twenty-four, and arranged in two series ; and it is generally found that the capitular radiations correspond in number with one series of the tentacles. Andres (p. 311) regards the species as synonymous with *Polythoa axinellæ*, Schmidt. This has since been described by Haddon and Shackleton (1891, p. 654), who make it the type species of the present genus. It will be found from the account here given, that the West Indian representative differs from the description of the European example, likewise commensal with a sponge, in many features both of form and anatomy.

The extraordinary abundance of the bright yellow pigment granules throughout the ectoderm and endoderm should be noted in the present species.

REFERENCES.

1707. SLOANE, SIR HANS :
  "A Voyage to the Islands of Madera, Barbados, Nieves, St. Christophers, and Jamaica. With the Natural History, etc., of the last of these Islands."—Two vols., London.

1786. ELLIS, J., AND SOLANDER, D. :
  "The Natural History of many curious and uncommon Zoophytes collected from various parts of the Globe."—London.

1817. CUVIER, G. C. L. D. :
  "Le Règne animal."—Paris.

1817. LESUEUR, C. A. :
  "Observations on several species of the genus Actinia. Illustrated by figures."—Jour. Acad. Nat. Sci., Philadelphia, vol. i. pp. 169-189.

1828. GRAY, J. E. :
  "Spicilegia Zoologica."—London.

1834. BLAINVILLE, H. M. de :
  "Manuel d'Actinologie ou de Zoophytologie." —Paris.

1850. DUCHASSAING, P. :
  "Animaux Radiaires des Antilles."—Paris.

1857. MILNE-EDWARDS, H. :
  "Histoire Naturelle des Coralliaires ou Polypes proprement dits."—Vol. i., Paris.

1860. GOSSE, P. H. :
  "Actinologia Britannica. A History of the British Sea-Anemones and Corals."—London.

1860. DUCHASSAING, P., ET MICHELOTTI, J. :
  "Mémoire sur les Coralliaires des Antilles."—Mem. Reale Accad. Sci. Turin, Ser. II., Tom. xix.

1864. VERRILL, A. E. :
  "Revision of the Polypi of the Eastern Coast of the United States."—Mem. Boston Soc. Nat. Hist., I. (Read 1862, pub. 1864.)

1866. DUCHASSAING, P., ET MICHELOTTI, J. :
  "Supplément au Mémoire sur les Coralliaires des Antilles." Mem. Reale Accad. Sci. Turin, Ser. II., Tom. xxiii.

1869. VERRILL, A. E. :
   " Notes on Radiata : Review of the Corals and Polyps of the West Coast of America."—Trans.
   Councct. Acad., vol. i., 1868-1870.

1882. HERTWIG, R. :
   " Report on the Actiniaria drodged by H.M.S. ' Challenger ' during the years 1873-1876."—
   Zoology, vol. vi.

1883. VERRILL, A. E. :
   " Report on the Anthozoa, and on some additional species dredged by the ' Blake ' in 1877-1879,
   and by the U. S. Fish Commission steamer ' Fish Hawk ' in 1880-1882."—Bull. Mus. Comd.
   Zool. Cambridge, Mass., vol. xi., No. 1.

1883. ANDRES, A. :
   " Le Attinie."—Atti. R. Accad. dei Lincei, ser. 3a, vol. xiv.   (Printed later in " Fauna u. Flora
   d. Golfes v. Neapel, ix. Leipzig," 1884.   I have quoted from this.)

1885. ERDMANN, A. :
   " Ueber einige neue Zooantheen.   Ein Beitrag zur anatomischen und systematischen Kenntniss
   der Actinien."—Jenaische Zeitschr. Naturwiss., Bd. xix.

1888. HERTWIG, R. :
   " Report on the Actiniaria dredged by H.M.S. ' Challenger ' during the years 1873-1876."—
   Supplement.

1889. M'MURRICH, J. P. :
   " The Actiniaria of the Bahama Islands, W. I."—Journ. Morphol., vol. iii., No. 1.

1889 a. M'MURRICH, J. P. :
   " A contribution to the Actinology of the Bermudas."—Proc. Acad. Nat. Sci. Philadelphia.

1891. HADDON, A. C., AND SHACKLETON, ALICE M.
   " A Revision of the British Actiniæ.   Part II.: The Zoanthew."—Trans. Roy. Dublin Soc.,
   vol. iv. (ser. II.)

1891 a. HADDON, A. C., AND SHACKLETON, ALICE M. :
   " Report on the Zoological collection made in Torres Straits by Prof. A. C. Haddon, 1888-1889.
   Actiniæ: 1.   Zoanthew."—Trans. Roy. Dublin Soc., vol. iv. (ser. II.)

1895. CARLGREN, O. :
   " Ueber die Guttung Gerardia, Lac.-Duth."   Öfversigt af Kongl. Vetenskaps-Akademiens
   Förhandlingar, No. 5.

1895. VON HEIDER, A. R. :
   " Zoanthus chierchiæ, n. sp."   Arbeit. Zool. Inst. Graz., v.

1896. M'MURRICH, J. P. :
   " Notes on some Actinians from the Bahama Islands, collected by the late Dr. J. I. Northrop."—
   Annals N. Y. Acad. Sci., ix.

1896. HADDON, A. C. AND DUERDEN, J. E. :
   " On some Actiniaria from Australia and other Districts."—Trans. Roy. Dublin Soc., vol. vi.
   (ser. II.)

EXPLANATION OF PLATE XVII.A.

PLATE XVII. A.

The dimensions represented are practically the same as in the living condition.

Trans. R. Dubl. Soc. Ser. II. Vol. VI.

Plate XVII.A.

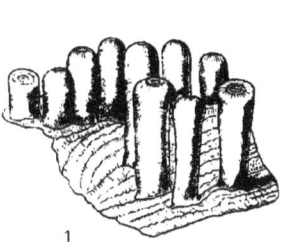

1

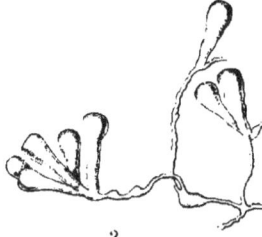

2

3

4

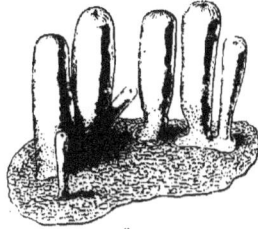

6

5

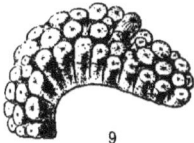

7

9

10

EXPLANATION OF PLATE XVIII. A.

3 L 2

PLATE XVIII. A.

Figure.

1. *Zoanthus Solanderi*, Les. (p. 335). Vertical section through a portion of the column-wall, × 50.

2. *Zoanthus flos-marinus*, Duch. and Michl. (p. 339). Vertical section through a portion of the column-wall, × 50.

3. *Zoanthus pulchellus* (Duch. and Michl.), (p. 341). Vertical section through a portion of the column-wall, × 200.

4. *Zoanthus pulchellus* (Duch. and Michl.), (p. 341). Transverse section through a portion of the column-wall and a fertile mesentery, × 50.

5. *Isaurus Duchassaingi* (Andres), (p. 346). Vertical section through the upper portion of the column-wall, × 25.

6. *Isaurus Duchassaingi* (Andres), (p. 346). Section through a portion of the mesoglœa of the body-wall, showing cells with peripheral granular protoplasm, × 250.

7. *Gemmaria variabilis*, n. sp. (p. 350). Vertical section through a portion of the column-wall, × 25.

8. *Gemmaria variabilis*, n. sp. (p. 350). Transverse section through a portion of the column-wall and a fertile mesentery, × 200.

9. *Gemmaria variabilis*, n. sp. (p. 350). Transverse section through a portion of the column-wall and a perfect mesentery in the region of the œsophagus, × 50.

10. *Gemmaria fusca*, n. sp. (p. 54). Vertical section through a portion of the column-wall, × 40.

Plate XVIII.A.

EXPLANATION OF PLATE XIX.

PLATE XIX.

Figure.

1. *Palythoa mammillosa* (Ell. and Sol.), (p. 359). Vertical section through a portion of the upper free part of a polyp, from which most of the incrustations have been dissolved, × 50.

2. *Palythoa mammillosa* (Ell. and Sol.), (p. 359). Transverse section through the œsophageal region (diagrammatic), × 35.

3. *Palythoa mammillosa* (Ell. and Sol.), (p. 359). Transverse section through the œsophageal region of a younger polyp (diagrammatic), × 35.

4. *Palythoa mammillosa* (Ell. and Sol.), (p. 359). Transverse section through a partially extended polyp, passing through a portion of the grooved part of the disc, × 50.

5. *Palythoa caribæa*, Duch. and Michl. (p. 365). Vertical section through a portion of the upper free part of a polyp, × 50.

6. *Palythoa caribæa*, Duch. and Michl. (p. 365). Transverse section through a fertile mesentery, × 50.

7. *Palythoa caribæa*, Duch. and Michl. (p. 365). Transverse section through the œsophageal region (diagrammatic), × 35.

Plate XIX

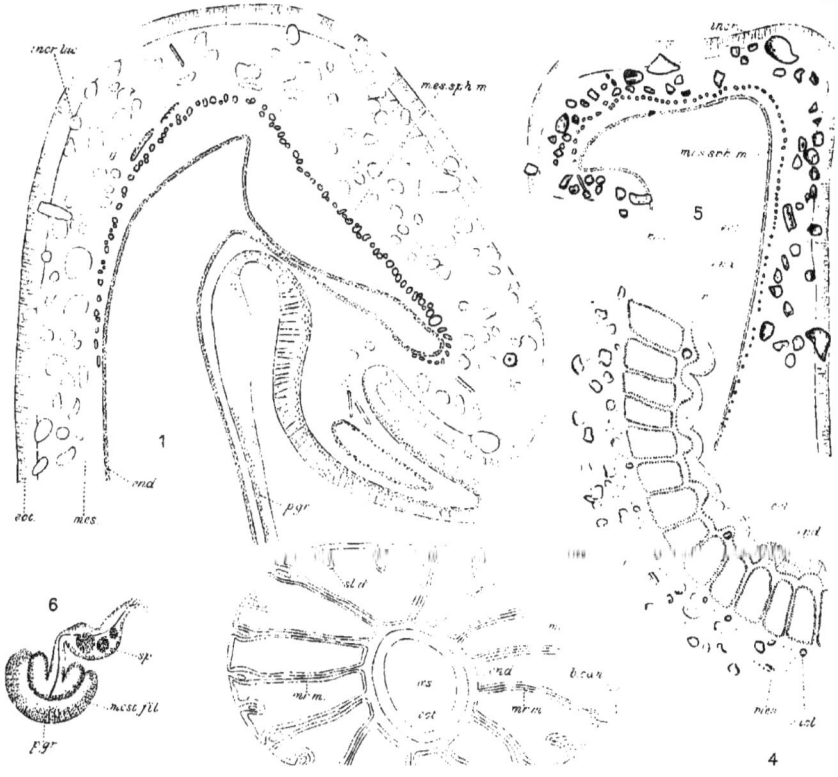

EXPLANATION OF PLATE XX.

PLATE XX.

Figure.

1. *Epizoanthus minutus*, n. sp. (p. 369). Vertical section through a portion of the column-wall, showing the sphincter muscle, tentacles (cut obliquely), disc, œsophageal wall, and mesentery, × 50. Polyp extended.

2. *Epizoanthus minutus*, n. sp. (p. 369). Transverse section through a portion of the column-wall and the grooved part of the disc in a partially extended polyp, showing the united bases of the tentacles, × 50.

3. *Epizoanthus minutus*, n. sp. (p. 369). Transverse section through a portion of a tentacle, × 200.

4. *Epizoanthus minutus*, n. sp. (p. 369). Transverse section through a portion of the column-wall, and the mesenteries below the œsophagus, × 50.

5. *Parazoanthus Swiftii* (Duch. and Michl.), (p. 372). Vertical section through a portion of the column-wall, base, and cœnenchyme, × 50.

6. *Parazoanthus Swiftii* (Duch. and Michl.), (p. 372). Transverse section through a portion of the column-wall and a fertile mesentery, × 280.

Plate XX.

TRANSACTIONS (SERIES II.).

VOLUME VI.

THE

SCIENTIFIC TRANSACTIONS

OF THE

ROYAL DUBLIN SOCIETY.

VOLUME VII.—(SERIES II.)

VI.

JAMAICAN ACTINIARIA. Part II.—STICHODACTYLINÆ AND ZOANTHEÆ.

By J. E. DUERDEN, Assoc. R. C. Sc. (Lond.), Curator of the Museum of the Institute

of Jamaica.

(Plates X. to XV.)

DUBLIN:
PUBLISHED BY THE ROYAL DUBLIN SOCIETY.
WILLIAMS AND NORGATE,
14, HENRIETTA STREET, COVENT GARDEN, LONDON,
20, SOUTH FREDERICK STREET, EDINBURGH; AND 7, BROAD STREET, OXFORD.
PRINTED AT THE UNIVERSITY PRESS, BY PONSONBY AND WELDRICK.
1900.

Price Three Shillings.

THE

SCIENTIFIC TRANSACTIONS

ROYAL DUBLIN SOCIETY.

VOLUME VII.—(SERIES II.)

VI.

JAMAICAN ACTINIARIA. Part II.—STICHODACTYLINÆ AND ZOANTHEÆ.
By J. E. DUERDEN, Assoc. R. C. Sc. (Lond.), Curator of the Museum of the Institute
of Jamaica.

(Plates X. to XV.)

DUBLIN:
PUBLISHED BY THE ROYAL DUBLIN SOCIETY.
WILLIAMS AND NORGATE,
14, HENRIETTA STREET, COVENT GARDEN, LONDON.
20, SOUTH FREDERICK STREET, EDINBURGH; AND 7, BROAD STREET, OXFORD.
PRINTED AT THE UNIVERSITY PRESS, BY PONSONBY AND WELDRICK.
1900.

VI.

JAMAICAN ACTINIARIA. Part II.—STICHODACTYLINÆ AND ZOANTHEÆ.

By J. E. DUERDEN, Assoc. R. C. Sc. (Lond.), Curator of the Museum of the Institute of Jamaica.

(PLATES X. TO XV.)

[Read DECEMBER 21, 1898.]

THE first instalment of this series (1898) was limited to the Zoantheæ occurring in the shallow waters around Jamaica, and ten species are described therein. As a result of trawling recently carried on over some of the deeper regions of the Caribbean Sea three other species of the same order, all belonging to the one genus Parazoanthus, have been procured. The account of these is included at the end of the present contribution.

This second communication describes seven species belonging to the mainly tropical order Stichodactylinæ. Several have already been anatomically studied and described by Professor McMurrich in his "Actiniaria of the Bahamas" (1889); and it is interesting to compare the many points of resemblance and difference as revealing the features of stability or of variation within the same form.

When the paper was practically completed I received the extremely valuable memoir by Professor A. C. Haddon, "The Actiniaria of Torres Straits" (1898), in which, besides describing fifty-five species from that particular locality, the author attempts a classificatory revision of the whole group.

Torres Straits has proved itself extremely rich in Stichodactylinæ. In so far as my results agree with those of Professor Haddon, I have been enabled to bring the present contribution into harmony with his conclusions.

Mention must also be made of Dr. Casimir R. Kwietniewski's "Actiniaria von Ambon und Thursday Island" (1898), an important paper also devoted to tropical Actiniæ, published while Haddon's memoir was going through the press, and therefore not referred to by him.

As it is not likely that Professor Haddon will, for some time, conduct other such elaborate investigations in this branch of zoology, he has most generously placed at my disposal his microscopic preparations of species already described by him, and has also lent me portions of his extensive collection of Actinozoan literature, favours specially appreciated in such an isolated position. For these, and for other encouraging help, I here beg to express my sincerest gratitude.

The present species are arranged as follows :—

Tribe.—HEXACTINIÆ, Hertwig.

Order.—Sticnodactylinæ, Andres.

Sub-order.—Heterodactylinæ, n. s.-o.

Family.—Phymantiidæ, Andres.

Genus.—**PHYMANTHUS**, Milne Edwards.
Phymanthus crucifer (Lesueur).

Family.—Rhodactidæ, Andres.

Genus.—**ACTINOTRYX**, Duchassaing amd Michelotti.
Actinotryx Sancti-Thomæ, Duchassaing & Michelotti.

Sub-order.—Homodactylinæ, n. s.-o.

Family.—Discosomidæ, Klunzinger.

Genus.—**RICORDEA**, Duchassaing & Michelotti.
Ricordea florida, Duchassaing & Michelotti.
Genus.—**STOICHACTIS**, Haddon.
Stoichactis helianthus (Ellis).
Genus.—**HOMOSTICHANTHUS**, n. g.
Homostichanthus anemone (Ellis).
Genus.—**ACTINOPORUS**, Duchassaing.
Actinoporus elegans, Duchassaing.

Family.—Corallimorphidæ, Hertwig.

Genus.—**CORYNACTIS**, Allman.
Corynactis myrcia (Duchassaing & Michelotti).

Tribe.—ZOANTHEÆ, Hertwig.

Family.—Zoantiidæ, Dana.

Sub-family.—Macrocneminæ, Haddon & Shackleton.

Genus.—**PARAZOANTHUS**, Haddon and Shackleton.
Parazoanthus tunicans, n. sp.
      ,,   **monostichus**, n. sp.
      ,,   **separatus**, n. sp.

Tribe.—HEXACTINIÆ, Hertwig, 1882.

Actiniaria with paired mesenteries. The mesenteries of each pair are provided with longitudinal muscular fibres on the faces turned towards each other, and with transverse muscles on the faces turned away from each other, except in the case of two (sometimes more, one or none) pairs—the directives—in which this arrangement of the musculature is reversed, so that the longitudinal muscles are on the faces which look away from each other. The number of pairs of perfect mesenteries is at least six, but may be eight, ten, or irregular, and they usually increase simultaneously in the same multiples.

The above definition is, in the main, that adopted by all writers since Professor R. Hertwig founded the tribe. As a result, however, of later investigations it has been found that exceptions may occur in almost every part of the original definition. Many forms are now known in which the hexameral symmetry is replaced by an octameral, decameral, or irregular arrangement; the directives may be absent, reduced to only one pair, or increased to more than two pairs. Even the increase in pairs of the cycles beyond the primary does not always proceed in regular multiples of the latter, or simultaneously. This is shown in *Ricordea florida* (Pl. XI., fig. 6 ; Pl. XII., fig. 1), where the pairs of the third cycle are developed very irregularly and never in proper alternation, *i.e.* double the number of the first or second cycle, as the rule of symmetry demands; the hexamerous plan is here likewise departed from.

Gonidial or œsophageal grooves, included by Hertwig in his definition, are now known to be so variable in number, or even to be absent in so many cases, that their inclusion in the tribal definition is of no importance. Dr. O. Carlgren (1893) adds that the column-wall and stomodæum are devoid of ectodermal longitudinal muscular and ganglionic layers, but, in the present paper, these are shown to occur in several species, and are already known for several others.

Order.—STICHODACTYLINÆ, Andres, 1883.

Hexactiniæ in which more than one tentacle may communicate with a mesenterial chamber. Usually a peripheral series of one or more cycles can be distinguished from an inner accessory series, the members of which are radially arranged or in groups, and are often of different form. Sphincter muscle either endodermal or absent.

The division of the tribe Hexactiniæ into the two orders, Actininæ (in which

X 2

only a single tentacle communicates with a mesenterial chamber) and Stichodactylinæ (in which more than one tentacle may communicate with a mesenterial
chamber), has the advantage of being founded upon an external character which
can be readily observed, and which must certainly be regarded as of some fundamental importance in Actinian morphology.

For better comparison of the tentacular relationships, I give a plan of a
portion of the disc in each case.

It is doubtful as to how far the Stichodactylinous condition is homologous
throughout the order, for important differences obtain in each of the seven species
to be described.

In the Phymanthidæ the marginal tentacles are in numerous, alternating,
entacmæous cycles, arranged exactly as are the tentacles in the Actininæ. The
inner, so-called tentacles are nothing more than mere discal tubercles, more or
less irregularly arranged, and histologically differ entirely from the outer
series. From the evidence afforded by its peripheral tentacles, I regard the
family as approaching the Actininæ more closely than any of the others.

The arrangement is somewhat similar in Actinotryx, but the marginal
tentacles are all in a single cycle, though they probably represent two or three
series for some reason not separated centripetally. The disc tentacles are
irregularly arranged with regard to the mesenterial chambers ; and their dendritic
form is perhaps but an exaggeration of the tubercular tentacle of Phymanthus.
The arrangement of the outer and inner groups in Actinotryx recalls that in
Corallimorphus, though the form of the tentacles presents a great contrast.

The case of Corynactis is otherwise. So far as my experience goes, no
distinction can be made between a peripheral and an inner series, though Haddon
(1898, p. 466) makes a generic character of such a separation. The tentacles in
each radial series follow one another in regular sequence, and afford the same
histological details, pointing to a common origin ; the relative sizes are, however,
the reverse of those in the Actininæ, *i.e.* the inner are the smaller, and the outer
the larger.

A somewhat similar arrangement holds in the genera Stoichactis and Ricordea.
The tentacles in the same radial row follow one another in a regular manner ; but
with regard to the conditions at the margin, however, the species vary. In
Stoichactis helianthus a single outermost cycle alternates with all the radial rows.
In *Ricordea florida*, on the other hand, the outermost cycle but one alternates with
all the rows within, and with the cycle peripheral to it ; these two marginal cycles
are somewhat larger, and are often of a colour distinct from that of the inner
tentacles.

Homostichanthus possesses about a dozen outer cycles of tentacles, often
distinguished from the inner series by the innermost cycle being differently

coloured from all the rest. They are all, however, on the same radii as the inner rows, which are not cyclic.

In Actinoporus the tentacles are simple or lobed vesicles, are practically all alike, and occupy all the radial divisions, two or more irregular rows communicating with the same mesenterial chamber.

Where the tentacles are so crowded, some of these relationships and distinctions are not easily recognized in contracted, preserved specimens. In living polyps, they can more readily be made out, often facilitated by colour distinctions.

I think it is desirable to have some division expressive of the similarity, or otherwise, of the tentacles in any genus, and therefore propose the following Sub-orders:—

HETERODACTYLINÆ.

Stichodactylinæ, in which the tentacles are of two forms, usually marginal and accessory, and separated by a naked portion of the disc. *Examples*—Phymanthus, Actinotryx, Rhodactis, Cryptodendron, Heterodactyla.

HOMODACTYLINÆ.

Stichodactylinæ, in which the tentacles are all of one kind, simple or complex, and usually follow one another in continuous rows. *Examples*—Discosoma, Ricordea, Stoichactis, Radianthus, Corynactis, Homostichanthus, Actinoporus.

Generally the more central tentacles are smaller or less complex than the more peripheral, but within the same species they are all formed on one plan.

Dr. Carlgren (1891, 1893) has erected the tribe Protantheæ, of which the most salient character is that the column-wall and stomodæum possess an ectodermal ganglion and longitudinal muscle layer. First formed to include the genera Gonactinia and Protanthea, in his later paper he embraces *Thaumactis medusoides*, Fowler, and the genera Corynactis and Corallimorphus, representing the families Thaumactinidæ and Corallimorphidæ respectively. Thaumactis has since been shown by Professor Haddon and myself (1896, p. 158) to be included in the family Aliciidæ, and I do not consider the presence of an ectodermal musculature of sufficient importance to warrant the separation of the Corallimorphidæ from its more natural place among the Stichodactylinæ.

A columnar and stomodæal ectodermal musculature and nerve layer are now known for many Hexactiniæ, the other characters of which indicate that they belong to totally different families. Professor McMurrich (1893, p. 143) refers to the probable occurrence of an ectodermal musculature in *Halcurias pilatus*; I have recorded it (1897) in two species of Bunodeopsis, and describe its presence in several of the forms to follow.   From his paper just received, I find that Professor Haddon and myself have independently come to the same conclusion with regard to the degree of importance to be attached to these histological details.   In the same contribution Haddon discusses at some length the points at issue between Carlgren's tribe PROTANTHEÆ and the PROTACTINIÆ of M°Murrich.

Regarding it as a relic from the ancestral Scyphistoma, Haddon (p. 411) considers that the ectodermal columnar and stomodæal musculature may persist amongst the lowest, *i.e.* the least specialized, members of various groups.   This view of its significance is further supported by the fact that it is often associated with a practically homogeneous mesoglœa, and sometimes with the absence of the " Flimmerstreifen " of the mesenterial filaments.   Such is the case in Corynactis and Ricordea, and partly so in Cerianthus, in each of which an ectodermal musculature occurs ; but in Phymanthus, and one or two others where the same structure is also developed, the mesoglœa is fibrous and includes numerous cells.

In a recent paper (1898), I have endeavoured to show that the combination of external and anatomical features met with in several of the Stichodactylinæ here described, are such as are also characteristic of the Madreporaria.

Sub-order.—HETERODACTYLINÆ, n. s.-o.

Family.—PHYMANTHIDÆ, Andres.

Thalassianthinœ,	.	. (pars), M. Edwards, 1857.
Phyllactiniœ,	.	. (pars), Klunzinger, 1877.
Phymanthidœ,	.	. Andres, 1883 ; M°Murrich, 1889 ; Kwietniewski, 1898 ; Haddon, 1898.

Stichodactylinæ, in which the tentacles are of two kinds : marginal tentacles arranged in several alternating entacmæous cycles, laterally tuberculiferous, or frondose ; inner tentacles radially or irregularly arranged, very small, tubercular or papilliform.

Genus.—**PHYMANTHUS**, Milne Edwards.*

Actinia,	.	. (pars), Lesueur, 1817.
Actinodendron,	.	. (pars), Ehrenberg, 1834.
Phymanthus,	.	. Milne Edwards, 1857 ; Klunzinger, 1877 ; Andres, 1883 ; M°Murrich, 1889 ; Kwietniewski, 1898 ; Haddon, 1898.

---

* While this contribution was going through the press I received from Prof. A. E. Verrill a copy of his paper : " Descriptions of new American Actinians, with critical notes on other species, I." Amer. Journ. Science, vol. vi., 1898, pp. 493-498.   In connexion with the genus of the species here referred

Phymanthidæ, in which the column is provided with longitudinal rows of verrucæ in its upper part, and terminated by a cycle of rounded acrorhagi. Sphincter muscle endodermal and very feeble, or absent. An ectodermal muscular and nervous layer are often present on the column-wall and stomodæum.

Professor Haddon (1898, p. 495) separates the genus Thelaceros, of Chalmers Mitchell (1890), from Phymanthus, solely on account of the absence of verrucæ on its column-wall. Kwietniewski (1898, p. 419), on the other hand, includes his species, *P. levis*, under the genus, although devoid of these structures, and places Thelaceros as a synonym of Phymanthus. It is clearly a matter of little moment which limitation is followed. As demonstrating the relationship of the two, it is important to note that Mitchell (p. 555) found a thick ectodermal musculature on the stomodæum of his Celebes representative.

<h3 style="text-align:center">Phymanthus crucifer (Lesueur).</h3>

(Pl. x., figs. 1 and 2 ; Pl. xi., figs. 1 and 2.)

Actinia crucifera, . . Lesueur, 1817, p. 174.
Cereus crucifer (Actinia), Duchassaing and Michelotti, 1866, p. 125; pl. vi., fig. 13.
Phymanthus cruciferus, . Andres, 1883, p. 501.
Phymanthus crucifer, . M'Murrich, 1889, p. 51, pl. ii., fig. 1 ; pl. iv., figs. 6–11.

With the exception of the oral disc, the polyps are usually buried in coral sand, or gravel ; the pedal disc is firmly adherent to rocks and stones, and adapts itself to the irregularities of their surface. In preserved specimens the base is flat, with coarse radial, and fine circular wrinklings, and is a little larger in diameter than the proximal region of the column.

The column is erect and smooth in the living condition, but exhibits coarse transverse and vertical wrinklings in contracted preserved specimens. When alive, the polyps are somewhat trumpet-shaped, expanding very slowly from the narrow region just above the limbus, until, in the upper region, they extend to two or three times their lower diameter. Distally the column is folded, and, *in situ*, this region rests upon the surface of the coral sand. The column is very thin-walled, and the lines of attachment of the mesenteries show through.

to, Prof. Verrill remarks (p. 496): "The generic name *Epicystis* [Ehr., Corall. rothen Meeres, p. 144, 1834] was proposed for the *Actinia crucifera* Les., *A. ultramarina* Les., and *A. granulifera* Les., the first being put in sect. *a*. Therefore, it is necessary to take the former as the type of the genus, which is evidently entirely distinct from the true *Phymanthus.*

Four to six very distinct, sucker-like verrucæ are developed in each row, and a few smaller examples may be continued below; the larger are capable of attaching pieces of fine gravel, fragments of shells, etc., to the column. The rows of verrucæ correspond with only certain of the mesenterial divisions as seen externally, and sometimes a single apical verruca may alternate with the principal rows.

A circle of prominent, rounded acrorhagi occurs at the apex of the column; these are double in number the rows of verrucæ, and alternate with the outermost row of tentacles. Sometimes a smaller acrorhagus alternates with the larger. A shallow fossa intervenes between the cycle of acrorhagi and the commencement of the tentacular region.

The peripheral tentacles are numerous, slightly entacmæous, shortly conical, and overhanging, the oral face being longer than the external. The number varies; the normal arrangement appears to be 6, 6, 12, 24, 48, &c.; one specimen bore 10, instead of 15, enclosed within the radii from two tentacles of the first cycle; the tentacles of many small polyps were counted in which the normal 96 were present, while in one specimen the irregular number 106 occured (Pl. x., fig. 1).

In the majority of polyps, the tentacles bear several transverse, opaque thickenings, most strongly developed along the oro-lateral area of the tentacles, where a distinct bilobation is often observable (Pl. x., fig. 2). Six or seven pairs of tubercles, arranged pinnately, may be present on the larger tentacles, diminishing a little in prominence both proximally and distally. The tentacles are smooth for some little distance from their origin, and remain so throughout their outer concave aspect.

Many polyps were procured wholly destitute of the thickenings, the tentacles being quite smooth, differently coloured, and presenting an entirely distinct appearance from the usual form. At first I had no hesitation in regarding these as a second species; but an acquaintance with scores of specimens, all living within the same area, revealed every stage in the presence or absence of the tubercles, some examples having only odd tentacles smooth, while others have only a few tuberculated.

The disc is very large, thin-walled, and, periphally, is thrown into 8–12 folds, and may overhang the column to a great extent; its middle region is flat, or may be slightly convex. The surface of the disc is characterized by the presence of small, wart-like projections, varying in size and arranged mostly radially; they correspond with the first and second cycles of tentacles, and sometimes with the lower orders. In large specimens, the tubercules may extend in great numbers over nearly the whole of the disc, even as far as the peristome, and vary considerably in number, size, and distance apart in each row. Before the peristome is reached they become more closely and irregularly arranged, and seem to correspond with all the mesenterial spaces (Pl. x., fig. 1).

Water was freely emitted through the tubercles when the animal was compressed in collecting, though it may be doubted if this occurs naturally. The peristome is slightly raised, and the gonidial grooves are well-marked, the two lips being thicker and lighter than the rest of the stomodæal walls.

The polyps possess very limited powers of retraction; the disc and tentacles are not infolded.

The colours are very variable, partly dependent upon the age of the specimens and whether the tentacles are smooth or bear thickened bands—but all gradations can be traced in an abundant series. The prevailing disc colours are brown and green, often iridescent, with opaque white spots and blotches; those of the column are scarlet and crimson on a white or cream ground.

The base is white, or may exhibit bright, radiating scarlet bands. The column is usually cream white, with irregular, longitudinal patches of scarlet; the verrucæ display a very pronounced deep crimson centre. The column-wall is light grey in young specimens, darker above. In these small forms the tentacles are also greyish, the thickenings appearing as transverse white bands. The concave, outer aspect of the tentacles is white, and a V-shaped white patch, with the angle open, occurs at the base. In some the thickenings are of the same brown colour as the tentacle itself. The smooth tentacles are brown or reddish brown, with light crimson tips, and three longitudinal lighter lines traverse the whole length.

The disc bears white, green, and blackish patches; a black or brown radial patch occurs at the inner aspect of the base of each of the primary tentacles, and two towards the base of the four next cycles. The peristome is usually iridescent green.

The dimensions are likewise very variable according to age. Many young specimens were collected, in which the length of the column was only 1·1 cm., and the diameter, 0·8 cm. When extended the disc is about 5·5 cm. across in large examples, but may be as much as 9 cm.; the length of the column is generally about 6 cm.; the diameter across the middle, 1·7 cm., and across the base, 2·5. The innermost tentacles measure 0·7 cm. in length.

The figure of this species given by Professor M'Murrich (1889, Pl. II., fig. 1), represents the more usual appearances of the Jamaican specimens, and I have not considered it necessary to add another. The marginal tentacles should, however, be compared with that on Pl. x. of the present paper.

ANATOMY AND HISTOLOGY.

The ectoderm of the base is a very deep, columnar epithelium, much broader than either of the two other layers, and, in sections, is thrown into strong folds, partly followed by the mesoglœa. Long, narrow, supporting cells are the chief constituents, but a few gland cells are also intermingled, though not by any means so thickly as in the ectoderm of the column-wall. Towards the mesogloea a fibrillar layer is very clearly displayed. No ectodermal muscle is present, or is only of the weakest character. The mesogloea is typical of its condition throughout the polyp, being very fibrous, and containing numerous cellular constituents. To its shrinkage is perhaps due, in some degree, the fine display of both the ectodermal and endodermal fibrillar layers which occur throughout all the polyps sectionized.

As shown in the ectoderm of Pl. xi., fig. 1, very fine fibrils, arranged in an almost parallel manner, extend from the delicate muscular layer, and afterwards unite to form an extremely thin layer. This latter usually appears as if made up of very delicate interlacing fibrillæ; and on its outer side another series of fibrils, irregularly arranged, are given off, and are connected with the cells of the ectoderm. For the sake of distinction I restrict the term nerve layer to the delicate, interlacing layer, and speak of that between it and the muscle layer as the fibrillar layer. Sometimes, as is represented in the endoderm of the same figure, the fibrils extending from the muscle fibres do not unite to form a definite nerve layer, but interlace and are reticular in section. What appear to be Ganglionic cells are distributed among the fibrillæ. Appearances similar to the above are also represented in the section through a portion of the tentacle and also of the gonidial groove of *Homostichanthus anemone* (Pl. xv., fig. 1 ; Pl. xiv., fig. 2), and can be made out in sections of most species.

The endoderm of the base contains many zooxanthellæ, which, along with the cell protoplasm and nuclei, are mostly concentrated towards the free border of the layer. A well-marked fibrillar stratum extends for some distance from the endodermal muscle, and the latter is arranged on fine mesogloeal plaitings.

The column-wall throughout is of only medium thickness, and becomes thinner towards the apex ; the mesogloea is nowhere much broader than the ectoderm or endoderm. The ectoderm is very deeply ridged in places, the elevations being partly followed by long processes of the mesogloea, often branching, and much longer than the whole thickness of the wall. The supporting cells are scarcely so long as at the base; but unicellular glands, some with granular contents and others quite clear, are abundant. Maceration preparations reveal the presence of numbers of small nematocysts. A very weak longitudinal ectodermal muscle is present, and a nervous layer is readily distinguishable in places. The endoderm is a deep layer containing zooxanthellæ; its constituent cells are considerably elevated between the mesenteries where these are closely arranged. Throughout the column the endodermal circular musculature is developed with exceptional uniformity, the mesogloea being finely plaited for its support, and a nervous layer is clearly seen in some parts (Pl. xi., fig. 1).

As already ascertained by M‘Murrich, there is no special concentration of the endodermal muscle to form a sphincter; indeed, the musculature, if anything, becomes more feeble towards the apex.

The verrucæ are readily distinguished from the rest of the column by the absence of gland cells from the ectoderm, and by the nuclear zone being broader and staining more deeply.

Sections through the acrorhagi reveal a thinner wall, an absence of gland cells, and small nematocysts.

In the tentacles, both the ectodermal and endodermal muscles are well-developed on mesoglœal plaitings. The nematocysts in the former layer are extremely small, and, in both, the fibrillar layer extends for some distance from the mesoglœa. Zooxanthellæ are abundant in the inner layer. Where the sections include one of the tentacular swellings, the enlargement is seen to be due to a slight increase in thickness of the endoderm, but more especially of the mesoglœa.

The disc is much like the tentacles in structure, but thinner-walled, and fewer nematocysts occur in the ectoderm. The mesoglœa is thrown into elongated branching folds to serve as an additional support to the endodermal circular muscle.

Sections through the wart-like projections reveal slight, hollow upgrowths of the disc; the mesoglœa and musculature become so attenuated as to be with difficulty recognizable, and the ectoderm is thinner than elsewhere, but apparently does not exhibit any new structural elements. Like M‘Murrich, I have been unable to determine if the tubercles are actually perforated, but the delicacy attained by both the mesoglœa and ectoderm at the apex is an indication that the production of a temporary opening by any pressure from within would be a matter of very little difficulty. It has already been noted that water may be omitted in the living condition. M‘Murrich (1889, pl. iv., fig. 11) gives a figure of a section through one of the disc tentacles.

The stomodæum is very elongate and oval-shaped in transverse sections, and extends almost across the cœlenteron, the pair of directives at each extremity being much shorter than the lateral complete mesenteries. In longitudinal sections, the stomodæum is comparatively short. Its walls are thin, and the ectoderm is thrown into irregular vertical folds, partly followed by the mesoglœa. The gonidial grooves are clearly indicated at each end, their walls unfolded, and not much thicker than the rest of the stomodæum. At each end the groove is prolonged for some distance below the lateral walls, the directives remaining attached all the way. All the other mesenteries, however, are very deeply concave along their free edge towards their attachment to the stomodæal wall, so that in transverse sections through this region a short portion of each mesentery still remains attached to the stomodæum and hangs freely from it, while the longer portion is connected with the column-wall and appears as a free mesentery. Histologically, the stomodæal ectoderm consists mainly of ciliated supporting cells and long, narrow gland cells with finely granular contents; the nuclei of the former give rise to a brightly-staining middle zone. A weak longitudinal ectodermal musculature and a nervous layer are distinctly recognizable. Nemato-cysts cannot be detected in sections. The mesoglœa is folded on its endodermal border for the support of the weak endodermal muscle. The endoderm itself is an extremely narrow layer, displaying nerve fibrillæ.

In polyps small enough to be sectionized as a whole, twelve pairs of perfect mesenteries are present, of which two pairs are directives. These constitute the first and second cycles; and it is usually found that the members of the second cycle become free in advance of the others; the directives, as already stated, extend much below the others, retaining their attachment to the prolonged gonidial grooves. A third cycle of twelve mesenteries alternates with the com-plete members, and extends some distance within the cœlenteron, while a fourth cycle of twenty-four mesenteries extends but a little beyond the column-wall. In dissections of large polyps, a fifth cycle of forty-eight pairs of mesenteries occurred, the fourth cycle now stretching well within the cœlenteron, and the third cycle becomes connected with the stomodæum for some distance. Trilobed mesenterial filaments are borne by the members of the first three cycles. The mesenteries are exceptionally narrow, and, as a whole, occupy but a small proportion of the cœlenteron. The perioral stomata are very pronounced, but the marginal stomata are small and not readily distinguished.

The retractor muscle is arranged on very narrow, bifurcating, mesoglœal pro-cesses and gives rise to a thickened band, distally situated at a considerable distance from the column-wall (Pl. xi., fig. 1), but nearer proximally. In sections it commences as a rounded or acute enlargement, and centrally diminishes gra-dually. The musculature beyond the region of the retractor is very feeble, except towards the origin of the mesentery in the column-wall, where, in the region below the stomodæum, a parieto-basilar muscle is well developed, and a strongly plaited pennon is present. The endodermal muscle is continued between the mesoglœa at the base of the mesentery, and that of the column-wall. The retractor is developed on all the cycles of mesenteries except the lowest; while in the proximal region a muscular layer is continued all round the smallest mesen-teries, the mesoglœa being folded. The oblique musculature is very feeble.

The mesenterial epithelium is narrow, except towards the origin of the mesen-teries, where it broadens, and what appears to be nerve fibrillæ become very obvious; the cell contents are here aggregated mostly peripherally (Pl. xi., fig 1).

For a short distance both above and below the inner termination of the stomodæum the mesenterial filaments are trilobed in transverse section, the middle lobe bearing the glandular streak or Nesseldrüsenstreif, and each lateral lobe a ciliated streak or Flimmerstreif (Pl. xi., fig. 2). I have elsewhere (1898, p. 644) suggested that it is scarcely correct to regard the term middle lobe as synonymous with the two first terms, nor the lateral lobes as synonymous with the second terms. The lobes are very complex in their structure, and different regions in each are marked out by very distinct histological characters.

In Phymanthus, as in practically all compound filaments, the apical region of the middle lobe can be sharply distinguished histologically from its lateral regions. The apex stains more deeply, owing to the presence of numerous ciliated supporting cells with oval nuclei, and usually contains intermingled granular gland cells and nematocysts. Beyond this the histological elements are not so closely aggregated; the cell nuclei are rounded, the ciliated layer is not so strong, nematocysts are absent, and zooxanthellæ usually occur. With very little alteration this tissue is continued for a short distance on to the lateral lobes, where it becomes replaced by another, which is at once distinguished from all other structures of the Actinian polyp by the brightly-staining, homogeneous character of its cell constituents. Both in sections and in macerations, these are seen to consist of ciliated supporting cells, the oval nucleus being arranged at different heights in the various cells, so as to produce a nucleated zone, extending practically across the thickness of the layer; neither gland cells nor nematocysts are ever mingled with the supporting cells, and the ciliation usually persists in preserved specimens, even when absent from all other regions of the polyp.

The hinder region of the filament, where it is connected with the mesentery, displays still another histological modification. From the lateral mesoglœal axis a fibrillar reticulation is developed, and extends nearly to the periphery of the layer, but there becomes more like the ordinary endoderm in character, possessing nuclei, gland cells, and, usually, zooxanthellæ. As this passes on to the mesentery, it becomes indistinguishable from the mesenterial endoderm. The mesoglœa of the axes supporting the middle and lateral lobes also becomes modified. It is broader than the mesenterial mesoglœa immediately behind, and oval or stellate cells are numerously developed, so that the whole structure stains very deeply, and stands out very prominently from the surrounding tissue. Similar details are afforded by the filaments of nearly all Actiniaria which possess a trifid mesoglœal axis (cf. Pl. xi., fig. 2 and Pl. xv., fig. 4, the latter a zoanthid).

It is usual to recognize in the trilobed mesenterial filament only the glandular streak and the ciliated streaks, and these terms have been generally employed as synonymous with the middle and lateral lobes, respectively. It seems desirable, however, to distinguish more closely the various regions exhibiting a different histological structure, and presumably performing a different function. The term *glandular streak*, and its German equivalents, Drüsenstreif or Nesseldrüsenstreif, I would restrict to the tissue at the apical region of the middle lobe; *ciliated streak*, or Flimmerstreifen, to the deeply-staining, strongly-ciliated region of the lateral lobes; *intermediate streak*, to the region on each side, partly developed on both the middle and lateral lobes, and separating the two zones already indicated; *reticular streak*, to the tissue along the basal region of the filament, the term most nearly expressing its character in microscopic sections.

When the ciliated streak disappears proximally, and the filament becomes what is known as simple, a trilobed outline in transverse section is nevertheless often preserved. Here the lateral enlargements, however, bear no relation except that of position, with the true trilobed filament. They are of the same significance as those to be described in connexion with *Corynactis myrcia* (Pl. xv., fig. 3), that is, each is simply a swelling of the ordinary mesenterial endoderm, there being little histological modification and no special supporting mesoglœal axis.

Histologically the whole of the terminal lobe in the simple filament appears to correspond with the apical tissue of the middle lobe in the trifoliate filament, that is, with the glandular streak as here restricted. Nematocysts generally become more numerous proximally, and the filaments are branched along with the free edge of the mesentery.

A peculiarity connected with the glandular streak in the present species is that the actual apex in some cases becomes deeply grooved, resembling that figured by the Hertwigs (1879, pl. v., fig. 14) for one of the lower mesenteries of *Sagartia parasitica*, and recalling the condition in the so-called "ectodermal filaments" of the Alcyonaria.

Ripe ova were present on mesenteries of the higher orders, in one specimen sectionized. McMurrich found all the mesenteries, even the directives, to be gonophoric. His Bahaman specimens were also hermaphrodite, but none of the Jamaican examples show such a combination of gonads, nor was this the case with the two species described by Kwietniewski (1898).

The species is found in considerable numbers, associated with Asteractis, in the crevices of coral rock, or on stones embedded in the coral sand, in the shallow waters on the south east-shore of Drunkenman Cay, outside Kingston Harbour.

They are seen with only the disc exposed, the overhanging portion of the column resting on the surface of the sand, while the base may be buried to a considerable depth.

The mottled, greyish colours of the younger forms harmonize with their surroundings, but the brighter colours of the adults offer a great contrast. It is among the abundance of specimens occurring at Drunkenman Cay, that the forms with smooth tentacles are to be procured. Large examples of the species are also met with at Port Antonio. All these possess the thickenings on the tentacles, and the disc tubercles are strongly developed.

It is evidently a common West Indian anemone, having been recorded from the following localities : attached to stones on the sand-banks of the island of Barbados (Lesueur) ; St. Thomas and Barbados (Duchassaing and Michelotti) ; usually fastened to blocks of coral rock in shallow water, Bahamas (M'Murrich).

Professor M'Murrich has already described the form in considerable detail. The salient features in which the Jamaican examples differ are the variability in colour, and the entire absence, in some instances, of the thickenings on the tentacles. The first mentioned character is especially noticeable in young specimens, these being largely a mottled grey and black, in strong contrast with the more brilliant colours of the large examples. *P. mucosus*, from the Australian seas, also displays somewhat similar colour variations.

The presence or absence of the thickenings on the tentacles would be worthy of at least specific distinction were it not that every gradation can be traced between the two extremes.

In respect to the tubercles on the tentacles the West Indian representative of the genus should be compared with the several species known from the Indo-Pacific region. The former never shows anything beyond simple or bilobed thickenings on the oro-lateral aspect of its tentacles, while the tentacles on *P. loligo*, Ehr., from the Red Sea, may bear pedunculate and branched outgrowths ; *P. mucosus*, H. & S., and *P. levis*, Kwiet., also carry slightly dendritic appendages.

Family.—RHODACTIDÆ, Andres.

Phyllactininæ (pars), . . Klunzinger, 1877.
Rhodactidæ, ... Andres, 1883 ; (pars) M'Murrich, 1889 ; Haddon, 1898.

Stichodactylinæ, in which the tentacles are of two forms, marginal tentacles of the ordinary form, arranged in a single cycle ; inner tentacles lobed or tuberculiform and irregularly arranged.

Under this family Professor M'Murrich includes the two West Indian species, *Actinotryx* (*Rhodactis*) *Sancti-Thomæ* and *Ricordea florida*. Owing to more recent researches, it seems to me imperative to remove Ricordea from this association and to assign it a place among the Discosomidæ, a position already hinted at both by MM. Duchassaing and Michelotti (1866, p. 122), and by Professor Verrill (1869, p. 462).

The form and arrangement of the tentacles must undoubtedly be the determining consideration in the classification of the order ; and in the two species mentioned, these bear no close relation one to the other. Ricordea agrees with the chief characteristic of the Discosomidæ in possessing tentacles all of one form and covering nearly the whole of the disc, while there is a marked difference, both in form and histology, between the marginal and discal tentacles in Actinotryx.

In important anatomical details, such as the deep stomodæal folds, absence of gonidial grooves, and ciliated streak, Actinotryx and Ricordea are related; but, excepting perhaps the last mentioned, evidently not much reliance can yet be placed upon these for indicating broader relationships.

It is a question of choosing between external characters, and internal anatomy and histology as the chief factors of relationship. In the order Stichodactylinæ, at any rate, I am of opinion that the best results will be attained by giving the greater prominence to the arrangement of the tentacles among the former. The frequent multi-oral condition of the disc occurring in both, must assuredly be looked upon as a specific peculiarity.

Haddon (1898, p. 477) includes the four genera, Rhodactis, Actinotryx, Ricordea, and Heteranthus in the family. The second genus was instituted by Duchassaing and Michelotti (1860, p. 45) for a West Indian form, which later was regarded by M°Murrich as allied to the *Rhodactis rhodostoma* of the Red Sea, and was therefore transferred to that genus. From details of this species supplied by Dr. Carlgren, Haddon (p. 477) considers that the two should remain as distinct genera, and this is the conclusion I have followed.

Genus.—ACTINOTRYX, Duchassaing & Michelotti.

Actinotryx, . . Duchassaing and Michelotti, 1860; Andres (Actinothrix), 1883 (pars); Haddon, 1898.
Rhodactis, . . M°Murrich, 1889.

Rhodactidæ, in which the column is smooth. Marginal tentacles short, in two or three series, but forming only one cycle; inner tentacles dendritic or lobed, partly separated into a middle discal group and a perioral group. Sphincter muscle feeble or absent. An ectodermal muscular layer on the column-wall and stomodæum. Stomodæum deeply furrowed. No gonidial grooves. Mesenterial filaments devoid of ciliated streak.

The history of the genus has just been given under the discussion on the family.

Actinotryx Sancti-Thomæ, Duch. & Mich.

(Pl. x., figs. 3–6; Pl. xi., figs. 3, 4; Pl. xii., fig. 3).

Actinotryx Sancti-Thomæ, Duchassaing & Michelotti, 1860, p. 45, pl. vii., fig. 2; 1866, p. 35.
Actinothrix Sancti-Thomæ, Andres, 1883, p. 509.
Rhodactis Sancti-Thomæ, M°Murrich, 1889, p. 42, pl. i., fig. 9, pl. iv., figs. 2, 3.

The base is very spreading, irregular in outline, and firmly fixed to coral rock, adapting itself to the irregularities of the surface; in diameter it is larger than the column. A brownish, cuticular layer occurs between the ectoderm of the base and the surface of attachment, and may either be detached with the polyp or remain adhering to the rock. The internal, radiating mesenterial lines are clearly indicated through the thin basal-wall.

The column is short, thin-walled, and semi-transparent, the mesenterial lines showing through; the surface is smooth, and thrown into delicate longitudinal ridges and furrows, and occasionally transverse wrinklings are indicated. The limbus is spreading and irregular, and the polyps are generally constricted a little above the middle. The upper region of the column and the disc usually overhang so as to hide the parts below. Individual polyps are generally oval in outline in transverse sections.

A single cycle of very short tentacles occurs at the apex of the column or margin of the disc, a longer tentacle alternating with a short one, or there may be two or three of the latter between two longer ones. Much irregularity occurs in this respect. In shape the tentacles may be shortly acuminate, or cylindrical and rounded at the tips. A small, nearly horizontal, tentaculiform outgrowth projects from near the base of many of the larger tentacles, and, occasionally, from the middle tentacle of an alternating three.

The disc is divisible into two regions: a peripheral, naked area, limited outwardly by the marginal cycle of tentacles; and a larger, inner area bearing the branching outgrowths described as tentacles. The peripheral portion of the disc is very thin, and sometimes stands as a distinct parapet or collar around the inner part; at other times it is entirely reflexed (Pl. x., fig. 3). The discal tentacles are very short, thick, columnar outgrowths which divide towards the apex into numerous— 6 to 12—small, tubercular, finger-shaped, or pointed processes. A few, irregularly arranged examples may occur near the mouth, or they may be practically absent from this region. The organs are capable of assuming the flaccid condition, or one of distension, independently of one another. They may be distributed over the whole disc except near the margin, but are always less numerous centrally. The distinction between the discal and the perioral series is not always apparent. Usually the tentacles appear irregularly disposed, but occasionally an arrangement in cycles, or, more often, in radial series, is evident. It is doubtful if more than two or three ever communicate with the same mesenterial chamber (Pl. x., figs. 3-6).

In the living condition the disc usually overhangs so as to hide the base and column; but the peripheral, naked portion of the disc may be vertically elevated as much as 4 mm. above the central part. In retraction the capitular part of the column closes over the rest of the disc as a thin, radiately marked, semi-transparent membrane, and all the marginal tentacles come together in the centre and close up the aperture.

The peristome is considerably elevated and usually elongated; like the rest of the disc it bears tuberculate bodies, which are here usually a little smaller. The mouth is circular in small forms, elongated in larger. The stomodæal wall is thrown into numerous, deep, much flattened folds, and is capable of considerable eversion. It is very sharply marked off from the disc. Twenty folds were counted in a very small specimen, and over thirty occurred in another; true gonidial grooves do not occur (aglyphic).

Polyps are occasionally come upon in which the disc bears two or more distinct mouths, without any indication of longitudinal fission.

The coloration varies much in detail in different polyps. The column-wall is light or dark brown, usually darker above, but lighter again towards the apex, the latter often showing an iridescent green tinge. The disc is an iridescent green, with brown, radiating lines separating it into narrow, radial areas; or, the general surface of the disc may be brown, and the radiating lines iridescent green.

Some of the radial bands may be nearly opaque white. The larger marginal tentacles are a faint blue below, brownish or rose-coloured distally; the smaller tentacles are brown. The disc tentacles are brown and iridescent green, or may be a deep blue purple. The stomodæal-walls are white.

The dimensions are very variable. The base may be 2·5 cm. across; the column 1·4 cm. in diameter, and 1 to 2 cm. high. The length of the larger marginal tentacles is 0·2 cm. One specimen was obtained with a disc 5·4 cm. in length, and length of mouth, 1·3 cm.

ANATOMY AND HISTOLOGY.

The ectoderm of the base is formed of high supporting cells, along with a few mucous cells. In sections the nuclei of the former give rise to a distinct narrow zone, situated a little nearer the mesoglœa. A very thick, brown, cuticular membrane, which readily strips off, is still present in some examples, and is continued for a short distance up the column. The mesoglœa, as is the case throughout the whole polyp, is very thin, and rather large ovate cells are sparsely scattered through it; a very weak musculature appears both on its ectodermal and endodermal borders. The endoderm is much narrower than the ectoderm, and is well supplied with gland cells, but devoid of zooxanthellæ.

The ectoderm of the column-wall is a high layer, and crowded with clear, unicellular, mucous glands, so that an outer zone appears colourless, the contents of the glands not staining. The cells are extremely delicate, and in sections are nearly always collapsed or shrunk; indeed the tissues as a whole are less resistant than any other form I have studied. This is probably a result of the excessive abundance of glandular cells in both ectoderm and endoderm. The nuclei of the ectodermal cells of the column-wall become aggregated towards the mesoglœa, and the muscular layer is distinctly recognizable where much shrinkage has not taken place. No nematocysts are present.

The mesoglœa is thin, but distally is formed into irregular plaits for the support of the endodermal muscle. On its ectodermal aspect it is thrown into the deep ridges with wide intervening furrows, already noticed among the external characters.

The endoderm bears numerous zooxanthellæ, especially above, and also scattered nematocysts. The cells are long and narrow, and the nuclei are arranged in a very narrow zone. The endodermal musculature commences a little above the base as a very thin, smooth layer; distally, however, it is more developed on slightly branching mesoglœal plaitings, being most concentrated just below the marginal tentacles (Pl. XI., fig. 3). This exaggeration should probably be regarded as a diffuse endodermal sphincter muscle, continuous on the one hand with that of the column-wall, and on the other with that over the smooth naked portion of the disc. It is obvious that no very precise distinction can be drawn in the early stages of development between a strong endodermal muscle and a concentration which may be spoken of as a diffuse endodermal sphincter; and, even in the same species, there is undoubtedly much variation in the extent of the mesoglœal foldings, according to the degree of retraction in which the polyp happened to be preserved.

The musculature appears to be a little better developed than was found by Prof. M⁰Murrich in his Bahaman specimens. In these the circular muscle of the column is described as being throughout exceedingly feebly developed. The extent of development here given was met with in two examples sectionized; but in a third the concentration of the muscle fibres and folding of the mesoglœa was observable to a much less degree, the condition evidently being about the same as that found by M⁰Murrich. The two first specimens were somewhat infolded in preservation, while the other was extended.

Professor M⁰Murrich (p. 44) notes as a peculiar feature of the ectoderm of the disc, marginal tentacles, and disc tentacles that nematocysts are entirely absent. I have made a careful investigation of this exceptional condition, and, so far as regards the results obtainable from sections, my experience agrees with that of M⁰Murrich, but on submitting the marginal tentacles to maceration I find ectodermal nematocysts very abundant at the tips. They are elongated, rounded at each end, and the spiral thread is not very distinct, while the interior presents a granular appearance. The endoderm also contains many nematocysts, but they are of a totally different character. Maceration of the ectoderm of the disc

Z 2

tentacles does not, however, yield any stinging cells. Gland cells are not developed in the marginal tentacles to the same extent as in the column-wall. The endodermal muscle is continued into the tentacles only as a very thin layer, the mesoglœa not being plaited; but, in the naked region of the disc, directly from the base of the tentacles, it is again strongly developed on mesoglœal plaitings, and extends the whole width of the naked area, practically disappearing again as the disc tentacles are reached. The ectoderm of the disc, though still containing gland cells, is narrow, but the endoderm becomes enormously thickened, and presents the appearance of an extremely loose vesicular tissue. In the disc tentacles, the ectoderm and mesoglœa are thin, but the large cavity is almost filled with the loose endodermal tissue, among which are numerous zooxanthellæ and medium-sized ovate nematocysts, with a loose thread thrown into four or five loops; an extraordinarily large stinging cyst is met with here and there. These latter are enormous, horn-coloured cysts when mature, and bear tubercular or spine-like outgrowths. They are by far the largest met with in any anemone described from this region (Pl. XI., fig. 4).

The constituent cells of the endoderm readily separate on maceration. The supporting cells are of the usual type, but longer, and the free extremity is ciliated; the small nucleus occurs about the middle of the length. In places where one or more zooxanthellæ are enclosed the cell is greatly swollen.

The stomodæum is round in transverse section, but short vertically, and its walls are very deeply folded in the latter direction. These foldings which are so marked a feature amongst the external characters are seen on anatomical examination to be elevations of the ectoderm followed by the mesoglœa, but not by the endoderm. They often branch even more than is shown in M'Murrich's figure (1889, Pl. IV., fig. 3). The ectoderm presents the usual characters, having a clear peripheral zone ciliated on the outside, a broad middle zone of oval deeply-staining nuclei, and a narrow fibrillar zone. Nematocysts of various sizes are present, including the large, colourless cysts with a spiral thread. A feeble longitudinal musculature can be detected. The endoderm is much like that of the column-wall, but deeper, and fewer zooxanthellæ and no nematocysts occur. The endodermal muscle is clearly distinguished, and rather strong, arranged on small mesoglœal plaitings.

As already remarked by Prof. M'Murrich the arrangement of the mesenteries is difficult to determine, and appears very irregular in the alternation of perfect and imperfect pairs, as does also the number of pairs. In addition to the complete mesenteries an imperfect series is well developed, extending some distance within the cœlenteron, and, in certain regions, a second incomplete order is also formed. The pairs are closely situated at about equal distances all the way round, the endocœles and exocœles being of nearly equal width; the endoderm of the column-wall within the mesenterial spaces is sometimes much elevated. The number of mesenteries is considerable, one small specimen possessed fourteen complete pairs, and another eighteen. The mesoglœa is very thin, and the vertical retractor muscles are extremely weak, so as to render it almost impossible to determine the directives. I have been able to definitely ascertain the presence of the latter in only one instance, but others may occur. The parieto-basilar muscle, though weak, is clearly distinguishable on each side. The endoderm is a thick, highly glandular layer, and medium-sized stinging cysts are abundant in places. In some instances the cœlenteron appears filled with the mucus extended from the endoderm cells.

The mesenterial filaments consist only of the middle glandular streak or Nesseldrüsenstreif, the lateral ciliated streaks or Flimmerstreifen being absent (Pl. XI., fig. 4). They are seen to originate in direct continuity with the ectoderm of the stomodæum. At first, owing to the strongly folded condition of the stomodæal ectoderm the mesenterial filament is irregular, or rather the appearance is presented as if a portion of the terminal region of the stomodæum were still connected with the free edge of each mesentery. It is only a little below the stomodæum that the ordinary rounded appearance is assumed, but the Nesseldrüsenstreif is never clearly marked off from the mesenterial epithelium (Pl. XI., fig. 4). The nematocysts in the upper region are elongated and narrow, but below they are much larger, oval, and strongly spinous. Different stages in the growth of the large tuberculated nematocysts can be distinguished, the contents of the younger staining deeply with carmine. The mature individuals present a peculiar appearance when a group is cut through transversely. Irregular spine-like projections extend all the way round the thick horn-coloured wall, the interior appears filled with some coagulated substance, and here and there a cut end of the thread is indicated. The spiral threads themselves are finely and regularly spirally marked. Towards the free end of the mesenteries in the lower regions the endoderm is often loaded with zooxanthellæ.

Of several specimens sectionized, only one contained reproductive organs, spermaria apparently in a dehiscing condition (Pl. XII., fig. 3). The reproductive cells are found in the interior of the mesoglœa, and in escaping break through the endodermal tissue. They occur in only a few of the mesenteries.

In three polyps dissected in the living condition, a few embryos were found. They are large, dark-green, ovoid bodies, slightly narrower at one end, and about 1 mm. in length. On cutting open the animal, they escaped freely into the water.

Prof. McMurrich (1891, p. 303) found that the polyps retain the embryos in the interior of the body until they are furnished with two or four perfect mesenteries. During the month of September I observed examples in process of parturition. The disc and upper part of the column were almost entirely infolded, and then rapidly extended, several opaque, yellowish-green embryos being extruded each time.

The species is found rather plentifully around the coral reefs of most of the Port Royal Cays, and very large examples were obtained around Maiden Cay, and also at Port Antonio. It was met with at New Providence, Bahamas, by Professor McMurrich, while Duchassaing and Michelotti collected their types at St. Thomas ; so that it probably occurs on the coral reefs throughout the Antillean area.

The polyps occur in water of from two to three fathoms, firmly attached to coral rock, and usually in company with living coral. Associations of several scores may occur, giving a carpet-like appearance to the sea-floor.

The body-wall and disc are very delicate ; after a little rough handling in collecting the mesenterial filaments readily protrude, especially through the disc, and an abundance of mucus is also given out. Such a protrusion of the mesenterial filaments through any part of the body-wall is rarely met with in Actiniæ, but is a usual occurrence among the corals.

The asexual reproduction by intracalycinal fission is the same as in *Ricordia florida*, except that one does not so often meet with individuals showing the multi-oral condition, fission and separation evidently taking place more readily. A very elongated example was procured having a small second mouth at one end, round which the disc tentacles had become closely aggregated, but the column-wall showed no sign of division.

The multiple arrangement presented by the tentaculate areas in *Actinotryx bryoides*, described by Professor Haddon from Torres Straits, is in marked contrast with their irregular disposition in the West Indian species, as also the seven or eight smaller peripheral tentacles alternating between two larger. The "one or two short knob-like tentacles on most of the crenulations of the parapet " are perhaps comparable with the horizontal outgrowths on some of the marginal tentacles in the present form.

Sub-order.—HOMODACTYLINÆ, n. s.-o.

Family.—DISCOSOMIDÆ, Klunzinger.

Discostominæ, .   Verrill, 1869.
Discosomidæ, . . . (pars), Klunzinger, 1877.
Discosomidæ, . .   Andres, 1883; McMurrich, 1889, 1893; Kwietniewski, 1897, 1898; Haddon, 1898.

Stichodactylinæ, in which the column-wall is smooth or provided with verrucæ towards its upper portion. Oral disc usually of large size and lobed. Tentacles numerous, and covering the greater portion of the surface of the disc; all short and of one form; either finger-shaped, knobbed, pointed, or vesicle-like; a peripheral series, arranged in cycles, is usually distinguishable from an inner series, arranged only in radial rows; one or more rows may communicate with the same mesenterial space. Mesenteries very numerous, many of which are perfect. Sphincter muscle present or absent.

As the Actiniaria of tropical regions are more studied, the genera embraced under this family become more and more numerous. In addition to the type genus Discosoma, the genera Radianthus, Stichodactis, and Helianthopsis, all erected by Kwietniewski (1898), are anatomically known; Haddon (1898) adds Discosomoides and Stoichactis; while in the present communication I increase the family by including within it the genera Actinoporus, Homostichanthus, and Ricordea.

From these, and from the definition given above, it will be seen that the family includes a very heterogeneous assemblage of forms, corresponding in this respect with the Sagartidæ among the Actininæ. The only constant feature appears to be that the tentacles are all of the same form in any one species, and cover the greater portion of the disc; but apparently in no two genera are the peripheral and the inner tentacles similarly related. It will probably be found advisable later to separate as sub-families forms in which only one row of tentacles communicates with a mesenterial chamber from those in which, as in Actinoporus, two or more rows may originate from the same mesenterial chamber.

Following the work of M'Murrich, Simon, and Kwietniewski upon various forms, Professor Haddon (1898, p. 469), in his recent paper, endeavours to introduce some order into the group, but significantly remarks: "This family will require a good deal of working at before it can be satisfactorily classified." He does not, however, attach that importance to the relationships of the peripheral to the inner tentacles, and of both to the mesenteric chambers, from which I am hopeful that much assistance can be obtained in arranging the numerous members of the family.

Genus.—**RICORDEA**, Duchassaing and Michelotti.

Ricordea, . . . Duchassaing and Michelotti, 1860; McMurrich, 1896; Haddon, 1898.

Heteranthus, . . . (Klunzinger, 1877), McMurrich, 1889.

Discosomidæ, in which the marginal tentacles are small, dicyclic, finger-shaped or slightly knobbed; inner tentacles a little smaller, of the same form, in single radial rows. Sphincter muscle absent. No gonidial grooves. An ectodermal longitudinal muscular layer on the column-wall, and on the stomodæal wall. Numerous perfect mesenteries; mesenterial filaments devoid of ciliated streak.

My reasons for transferring this genus from the family Rhodactidæ, where it has usually been placed, to the family Discosomidæ, are given under the discussion on the former family.

Succinctly, the history of the genus is as follows : MM. Duchassaing and Michelotti (1860) first established Ricordea for the species about to be noticed ; in 1877, Klunzinger erected the genus Heteranthus for a very similar form, *H. verruculatus*, from the Red Sea ; Andres associated this latter species with the genus Actinothrix, which Duchassaing and Michelotti had erected for the form here described as *Actinotryx Sancti-Thomæ*, placing *Ricordea florida* among the "species incertæ sedis." McMurrich (1889), considering Duchassaing and Michelotti's definition of their genus Ricordea to be only specific, not generic, in character, disregarded it in favour of Klunzinger's Heteranthus ; in a later paper McMurrich (1896) returns to the original term Ricordea.

As at present known the genus includes with certainty only the one species, *R. florida*. The precise position of *Heteranthus verruculatus* cannot be established until an anatomical examination of it has been made. Haddon (1898, p. 481) suggests that two or three other species, described under different generic terms, may also be included along with *R. florida*.

Ricordea is undoubtedly closely allied to the two genera Discosomoides and Discosoma, as these are defined by Haddon (p. 470) from Simon's researches. The three agree in having a smooth column-wall, tuberculiform or papilliform tentacles, no gonidial grooves, numerous perfect mesenteries, and sphincter muscle absent or weak. Both *R. florida* and *D. nummiforme* are, in addition, characterized by the clearness of their mesogloea, and the exceptional structure of their mesenterial filaments.

Ricordea florida.—(Pl. x., fig. 7 ; Pl. xi., figs. 5, 6 ; Pl. xii., figs. 1, 2 ; Pl. xiii., fig. 1), Duchassaing and Michelotti.

Ricordea florida, .   . Duchassaing and Michelotti, 1860, p. 42, pl. vi.,
                                    fig. 11 ; 1866, p. 122 ; Andres, 1883, p. 572 ;
                                    McMurrich, 1896, p. 188.
Heteranthus floridus,   . McMurrich, 1889, p. 47, pl. i., fig. 10 ; pl. iv.,
                                    figs. 4–5.

The base is usually smaller in diameter than the column, and adheres so closely to the irregular surfaces of coral rock as to render almost impossible the removal of the polyp without injury or the separation of fragments of the rock.

The column is short and variously outlined. Simple forms are cylindrical, but compound examples are elongated laterally and sinuous above ; the limbus is generally irregular, and often of less diameter than the more distal part of the column. The column-wall is smooth and delicate, and the insertion of the mesenteries shows through. The wall is finely ridged in the living condition; while, in preserved specimens, it is thrown into rather deep, close, zigzag, vertical striæ, but no verrucæ occur.

The tentacles are short, knobbed or rounded at the apex, and arranged in two series: a marginal cyclic group, and an inner radiating group. The former are dicyclic and entacmæous; the latter extend for various distances from near the margin towards the centre, diminishing slightly in size centripetally. In places, a distinct serial arrangement of the radial tentacles is exhibited, those of the first order reaching as far as the peristome, the second and alternating order a little shorter, and a third and fourth still more so. The shortest series contains only two or three tentacles in each row. This regular and evidently normal arrangement is departed from in other parts of the disc. About 60 rows were counted in one specimen, but the number varies with the size of the polyp. Each marginal tentacle consists of a short stalk narrowing a little above, and terminating bluntly or in a slight knob. Professor McMurrich describes and figures the marginal tentacles of the Bahaman specimens as conical.

The inner tentacles are on the same radii as the outer cycle of marginal tentacles, thus alternating with the members of the first marginal cycle. They vary in length from 0·2 cm. to mere tuberculiform processes, the outer being the larger. Like those of the margin they terminate in a rounded or slightly knobbed manner, and the members of each radial row are in close contiguity peripherally, but become more distant one from the other centrally. Only a few of the radiating rows reach the peristome, and the tentacles here are a little larger than those for some distance behind (Pl. x., fig. 7; Pl. xi., fig. 5).

The disc is sinuous, and usually elongated or irregular in outline. It is often reflexed at the margin, and so thin-walled that in some the movements of the internal larvæ could be distinguished. The peristome is round and considerably elevated, ending sharply at the oval mouth. The stomodæal walls show about twelve very deep, flattened folds, but no gonidial grooves are indicated.

The number of oral apertures is inconstant. Probably the majority of Jamaican examples have more than one. A specimen was come upon which possessed seven mouths of different sizes. Duchassaing and Michelotti regarded five as normal when the development is complete. An example bearing three apertures was met with in the act of vertical fission, the elongated columnar constriction being nearly broken down; others were collected, joined only by a thin basal membrane. The disc and tentacles can be completely infolded, so that no part of them is visible.

The base is white, the proximal region of the column is flesh-coloured, the distal a very dark brown or may be bluish towards the margin. The marginal tentacles have a green, blue, or brown stem, and the knob green or yellowish-green, or white in one variety. Often there is a marked difference in colour between the dicyclic marginal tentacles and the inner radiating group. The disc is richly coloured; green or blue along the radii, but towards the inner naked area, dark purplish brown. The peristome is bright green; the walls of the stomodæum are white or light green.

Specimens were procured at Port Antonio in which the tip of the disc tentacles was coloured bright orange red. They were associated with other polyps of the more usual colours, to which they presented a marked contrast. Duchassaing and Michelotti also refer to a similar variety.

Around the Port Royal Cays two other slight colour varieties occur, also in close association; in one green and purplish brown predominate, and in the other light green or grey. They are readily distinguishable when seen *in situ* in patches of considerable size, the members of any one patch being alike, much in the same way as has often been described for groups of Corynactis.

The dimensions are very variable. The height of the column in extension may vary from 1·1 cm. to 2·7 cm., while the diameter may be 1·7 cm. The greater length of the disc of a specimen with four mouths, and also of one with a single mouth, was 3·5 cm. ; another bearing two apertures was 5·5 cm. in extension, with a diameter of 3·8 cm., and a length of 4 cm. in contraction.

The length of the marginal tentacles of the inner cycle is 0·5 cm.

ANATOMY AND HISTOLOGY.

In endeavouring to remove the animal from the rock to which it is attached, the base often becomes partly destroyed. Where perfect, however, the basal ectoderm is seen to be formed of elongated columnar cells, which are mostly large, unicellular glands, aggregated, along with the narrow supporting cells, around fine mesoglœal processes. A thick cuticular membrane, apparently formed of coagulated mucus, is present in some specimens between the ectoderm and the foreign object to which the polyp is attached.

The mesoglœa of the base, like that throughout the whole polyp, is a thin clear layer, and comparatively few cells are included within it. It stains slightly with borax carmine. In regard to the abundance of the mesoglœal cells, the species is intermediate between their practical absence in Corynactis and the great quantity in most Actiniæ. In the upper part of the polyp the layer is almost completely homogeneous, an included cell occurring but rarely. In these places it is indistinguishable from the mesoglœa of Corynactis and the Madreporaria. The endoderm of the base presents no important characteristics, and is practically devoid of zooxanthellæ.

As noticed among the external characters, the ectoderm of the column-wall is thrown into fine ridges and grooves. In section the former are shown to be supported by comparatively long mesoglœal processes, which may even become branched (Pl. xi., fig. 6). Numerous clear gland cells are present in the layer, along with fewer granular gland cells ; many of the former are fixed with the mucus in the act of streaming out. A very distinct though weak ectodermal musculature is seen in transverse sections, more readily noticeable on the mesoglœal processes. The mesoglœa is thin for its whole length, and becomes even more so towards the apex. The endodermal muscle is recognizable throughout the extent of the column as a feeble layer, but the fibres become a little stronger towards the apex, though no concentration which can be regarded as a special sphincter muscle takes place. The mesoglœa is thrown into very slight folds for its support, and very fine fibrils pass from these towards the free surface of the endoderm, where the nuclei and zooxanthellæ are mostly concentrated. The commensal algæ are much more abundant in the distal regions of the polyps than proximally.

The marginal and the disc tentacles display a similar structure. For the greater part of the length of the stem the ectoderm contains many gland cells, and its histology closely resembles that of the column-wall. Towards the free termination an important modification take place ; the majority of the cells are no longer broad and glandular, but elongated, narrow, and closely aggregated, while deeply-staining nuclei are abundant. A peripheral zone is made up almost entirely of large elongated cnidocysts, with a fine spiral thread inside. In the deeper parts of the layer others are seen in different stages of development, and very distinct fibrillar and nervous layers occur just outside the ectodermal muscle. The ectodermal and endodermal musculature are both very feeble.

Professor McMurrich (p. 48) found the tentacles in the Bahaman examples characterized by the total absence of nematocysts, a condition at variance with the Jamaican representatives, where both the marginal and disc tentacles are crowded with nematocysts around their extremity, rather more so in the marginal than in the inner tentacles. In a very young specimen sectionized, however, I was unable to discover any, even on maceration.

The mesoglœa of the tentacles is extremely thin, except proximally, where it becomes broader and almost homogeneous. The ectodermal muscle is rather strong around the base of the tentacles, but weakens distally. Numerous zooxanthellæ are present in the thickened endoderm.

The ectoderm of the disc contains very many, clear gland cells, and a few narrow nematocysts, and, in places, large, granular gland cells ; endodermal and ectodermal musculatures on fine mesoglœal processes are also clearly indicated. Zooxanthellæ are abundant in the discal endoderm, while few nuclei are to be seen in the mesoglœa.

2 A 2

The stomodæal ectoderm is deeply folded, a salient feature being the comparatively large mesoglœal processes, resembling those of the column-wall, which support the folds. As shown in the section figured (Pl. xi., fig. 6), these bear no relation to the attachment of the mesenteries on the inner side. There is no modification in structure indicating gonidial grooves. Large nematocysts, such as are met with in the mesenterial filaments, occur sparingly in the ectoderm, and a feeble ectodermal musculature can be discerned.

The mesenteries are very irregular in their development, but the hexameral condition is evidently the normal; one or more incomplete pairs may occur in the exocœles between the perfect pairs, or a pair may consist of one perfect and one imperfect mesentery. This irregular arrangement is consonant with what has been already noted for the tentacles, and is likewise probably connected with the usual method of reproduction by fission.

At least three distinct orders are indicated in most polyps, though many of the members of the third order may be wanting. The section (Pl. xi., fig. 6), of a young specimen shows thirty-six mesenteries of varying degrees of development; of these only thirteen reach the stomodæum. Another presented twenty-four pairs, of which eight pairs were perfect, though in two pairs one mesentery in each fell short of the stomodæum. A very small example sectionized exhibited only seven complete mesenteries, all arising from a region embracing little more than one-half of the circumference of the column-wall, while those arising from the remainder were all incomplete. The mesenteries become very numerous and closely arranged in large specimens.

Another polyp, sectionized later, is diagrammatically represented in transverse section in fig. 1, Pl. xii.; two pairs of directives occur, with three pairs of complete mesenteries on one side and four on the other. The members of the third order are present in some of the exocœles, but never in two pairs, one on each side of the pairs of the second order, as in the case where the cycles are developed regularly.

The retractor muscles, though feeble, are sufficiently well developed to allow of the arrangement in pairs being easily followed. No directives were distinguishable in two young specimens, but in another two pairs occurred. The parieto-basilar muscles are very distinct on each face, but the mesoglœa is smooth, and affords no indication of any basal pennon. The retractor muscles are supported, in places, on rather considerable plaitings of the mesoglœa, but the distribution of the plaitings is very irregular, and rarely presents the same appearance on any two mesenteries. In some cases they may form two or three thickened vertical bands. In the mesentery represented in fig. 2, Pl. xii., only one of these was present. Beyond the retractor region, the mesoglœa becomes extremely thin, sometimes appearing to originate from the side of the thicker part, instead of being a continuation of it. The retractor muscle extends over the whole face of the mesentery, and similarly with the oblique muscle on the other face.

The mesenterial endoderm is a rather broad layer, and, in the upper region, its cells contain numerous zooxanthellæ. Clear mucus is sometimes seen in the act of being extruded from many of the cells, and an occasional granular gland cell is present.

The mesenterial filaments have only the central portion or Nesseldrüsenstreif developed throughout their extent. This bears very large oval nematocysts with the spiral thread somewhat loosely arranged. They are located in the deeper parts of the filaments, and a narrower kind occurs at the margin. The filaments can be traced in connexion with the stomodæal ectoderm, and are nowhere very sharply marked off from the mesenterial epithelium (Pl. xiii., fig. 1).

No reproductive organs were present in the half dozen examples sectionized.

The species is found very abundantly, in water of three or four feet, growing on the coral rock around all the Port Royal Cays associated with *Actinotrix Sancti-Thomæ*; also at Laughlands, St. Ann, and at Port Antonio. Duchassaing and Michelotti record it from St. Thomas, and M°Murrich from the Bahamas.

The polyps are always aggregated in patches, often several feet across, as a result of their usual method of reproduction by fission. They display but little activity in opening and closing, the extended condition being by far the more usual. An excessive amount of clear mucus is given out on handling, rendering it very difficult to remove them from their attachment, and interfering somewhat with their proper preservation.

Genus.—STOICHACTIS, Haddon.

Discosoma, . . . M°Murrich, 1889, 1893; Kwietniewski (pars), 1898.
Stoichactis, . . . Haddon, 1898.

Discosomidæ, usually of large size; column smooth below, and with verrucæ above. Tentacles vary in form, from moderately short and subulate, to short and blunt, and even to quite small and capitate; not more than one row communicates with a mesenterial chamber. Sphincter muscle strong and circumscribed. Generally two gonidial grooves.

Consequent upon the researches of Dr. J. A. Simon (1892), on the type species of Discosoma—*D. nummiforme*, it was clear that some of the species included under the genus would have to be separated. For forms similar to the West Indian Discosomid, which M°Murrich first anatomically investigated, Haddon erects the above genus and includes, in addition, two Australian representatives—*S. Kenti*

162   J. E. DUERDEN—*Jamaican Actiniaria :*

(H. & S.) [the type], and *S. Haddoni* (S.-Kent), and also *D. Fuegiensis* (Dana). The sphincter muscle is remarkably similar in all four.

D. tuberculata, Kwiet., should also probably be transferred to Stoichactis, but not *D. amboncnsis*, Kwiet., in which the tentacles are placed in radial groups, so that more than one row communicates with a mesenterial chamber.

Whether, as in the form I identify as *D. helianthus* (Ellis), a single marginal cycle of exocœlic tentacles alternating with all the endocœlic radiating rows will be found in other representatives of the genus remains to be seen. Until this is ascertained it seems doubtful if the character should be assigned generic rank, and I have therefore omitted it.

Stoichactis helianthus (Ellis).

(Pl. XI., fig. 7; Pl. XIV., fig. 1.)

Actinia helianthus,	. . .	Ellis, 1767, p. 436, pl. xiii., figs. 6, 7; Ellis and Solander, 1786, p. 6.
Hydra helianthus,	. . .	Gmelin, 1788, p. 3869.
Discosoma helianthus,	. .	Milne-Edwards, 1857, p. 256; Duchassaing and Michelotti, 1866, p. 122; Andres, 1883, p. 493.
Discosoma anemone,	. . .	McMurrich, 1889, p. 37, pl. i., fig. 8; pl. iii., figs. 15–16, pl. iv., fig. 1.

The base is a little larger in diameter than the lower part of the column; usually it is firmly adherent to the surface of rocks, or may be buried in the sand. It adapts itself to the irregularities of any object to which it is attached, and is generally deeply wrinkled in consequence; preserved examples show concentric and radiating ridges and furrows.

The column is short and salver-shaped, narrowing a little above the base, and then expanding enormously in a crateriform manner, so as to completely overhang and hide the basal part. Usually the column is only partly embedded in sand, the overhanging distal region being free and resting on the surface. Its walls are somewhat thick, but slightly transparent; the surface is smooth, and grooved in correspondence with the attachment of the mesenteries. Distally vertical rows of oval green verrucæ occur, but they are evidently incapable of attaching foreign particles to the column. The apex of the column, corresponding with each mesenterial space, is slightly rounded, but is not modified to form an acrorhagus. A well-marked fossa occurs between the apex of the column and the base of the outermost row of tentacles.

The disc is greatly expanded, but remains flat, never being thrown into folds as in the next species. By far the greater part of it is covered with radiating tentacular rows of various lengths; the central, naked area is smooth, and the peristome somewhat elevated. Towards the periphery the tentacles are so close that the actual surface of the disc can scarcely be seen; centrally, as the rows begin to cease, the disc itself is more exposed. In a large specimen, 160 rows were counted near the periphery, the longest row containing 24 tentacles, all of which communicate with the same mesenterial space; interspaces may occur at almost any point showing where tentacles have failed to develop. Usually no serial order in the lengths of the various rows is apparent, though in young examples an arrangement in three or four orders can sometimes be made out. In larger specimens, all kinds of irregularities in the way of omissions may occur; a bifid example is occasionally come upon, and small tentacles are seen in process of development all round the margin.

A single outermost cycle alternates with all the radial rows (Pl. xi., fig. 7).

Viewed from within, in dissections, the rows of tentacles are seen to communicate with the endocœlic chambers by large, closely arranged, circular apertures; the outermost cycle is exocœlic in position. The tentacles are short and digitiform, but vary a little in shape and size, according to the amount of distension; sometimes they are quite collapsed. In the preserved condition the tentacles of some polyps retain the finger-shape in all, while, in most, they become short and vesicle-like, a denser apical area denoting the extent of the distribution of the nematocysts. The surface of the tentacles may be very finely fluted, from apex to base, in preserved specimens, and the discal ectoderm forms still finer ridges and furrows. The mouth is large and oval; two gonidial grooves are always present, readily distinguished by their thick lips. In large polyps three grooves are sometimes met with.

The base is white or cream-coloured; the column white or cream below, and a little darker above. Sometimes large, irregular, green patches may occur on the column and distally irregular vertical rows of small, oval-shaped, green areas represent the verrucæ, the number and closeness varying in the same specimen in different rows. The disc may be a lighter or darker olive brown, and the tentacles are the same, but irregular patches of different intensity are usually exhibited.

The peristome is a brownish yellow, the lips a rich yellow, the stomodæal wall white.

The waters at Port Antonio contain a remarkable colour variety. The entire column and disc, with the exception of the green verrucæ and a slight brown tint on the peristome, are colourless and perfectly transparent. The tentacles, on the other hand, are a clear, delicate, sulphur yellow. It is scarcely possible to imagine a form differing more in colour from the ordinary condition; seen on the dark sea-floor, they are very attractive objects. Odd brown tentacles may be scattered among the yellow ones, and one or two examples showed considerable areas of the disc with the usual colours, demonstrating that we are dealing with a mere colour variety, between which and the normal every gradation may occur.

The diameter of the base is about 5 cm., and the height of the column 4 cm. The diameter of the disc is 10 to 12 cm., or may be even more. The tentacles are about 0·6 cm. in length, and vary but little in different regions of the disc. They are often largest in diameter at the tips, where they may measure 0·2 cm. across. The diameter of the naked part of the disc across the mouth is 2·5 cm.

Prof. M'Murrich's figure (1889, Pl. I., fig. 8, *Discosoma anemone*) represents the usual appearance of the Jamaican specimens.

ANATOMY AND HISTOLOGY.

The column-wall is of only moderate thickness, the mesogloea being often narrower in section than the ectoderm. The latter is deeply folded, the mesogloea partly following. In the ectoderm the nuclei of the supporting cells are distributed with considerable uniformity in sections, not limited to a zone as is generally the case. Very numerous, long, granular gland cells are included among the supporting cells. There is no trace of any ectodermal musculature. The mesogloea shows a delicate, fibrous structure, and numerous included cells. On its endodermal border it presents narrow, slightly branching plaits for the support of the circular musculature, and very fine fibrils pass into the denser peripheral part of the endoderm, in some places giving rise to a distinct nerve layer. Ganglionic cells are recognizable between the muscular and nervous layers. The endoderm is much thinner than the two other layers, and contains many zooxanthellæ and granular gland cells.

Where sections pass through verrucæ, the ectoderm undergoes certain modifications : gland cells are absent, and the region stains more densely than the ordinary ectoderm. Very delicate processes, like cnidocils, also appear on the surface, and the remains of the layer of cilia are more obvious than elsewhere.

The sphincter muscle is a strong, circumscribed, endodermal representative. It is recognized as a large outgrowth from the column-wall, a little below the outermost cycle of tentacles, and is made up of several lobes. The pedicle is broad and short, and a narrow mesogloeal axis extends nearly the whole length. From this axis delicate processes are given off—sometimes on one side, sometimes on the other, or on both together—for the support of the musculature. The lobes are so deeply separated that often a portion of the coelenteron is enclosed in sections. The surrounding endodermal layer resembles that of the column-wall, and contains numbers of gland-cells and zooxanthellæ. Owing to the lobed character, the appearance presented by the muscle varies in different sections. As indicating the possible amount of this variation, the figure given by M'Murrich (1889, pl. III., fig. 15) should be compared with that on (Plate XIV., fig. 1).

The tentacles appear as simple outgrowths of the disc, no special sphincter being developed at their origin. The three layers are about equal in thickness, and may be a little folded in preserved material. Nematocysts are only borne towards the apex of the tentacles, but there is little or no enlargement distinguishing the capitulum from the stem. The nematocysts are rather long and very narrow, and the spiral thread inside is easily recognized. They are closely packed in a peripheral zone; below this is a broad nuclear zone; then a clearly defined thin nervous layer; and, lastly, the longitudinal ectodermal muscle on fine mesoglœal plaitings. The endoderm is crowded with zooxanthellæ and gland cells; the granular contents of the latter are in many cases in the act of being extruded into the tentacular cavity. There is only the merest trace of an endodermal musculature; endodermal nerve fibrillæ are distinguishable, but do not unite into a distinct layer, as in the ectoderm.

The ectoderm of the disc is devoid of cnidoysts, but contains numerous glandular cells with granular contents. A weak, radial, ectodermal musculature occurs, and the circular endodermal muscle is more strongly developed than in the tentacles, the mesoglœa being deeply plaited; the nerve fibrillæ are clearly seen in places, united into an extremely thin layer some distance from the muscle layer. Gland cells are abundant in the endoderm.

In a dissection of a small specimen, through the middle of the stomodæal region, twelve pairs of perfect mesenteries occurred, of which two pairs were directives; an alternating cycle of twelve pairs extended about half-way towards the stomodæum; and, of the third cycle, made up of twenty-four pairs, some extended only just beyond the column-wall, while others were larger. In other and larger specimens numerous irregularities were presented, pairs belonging to any of the cycles being missing or present in excess. The number of perfect mesenteries in these becomes very considerable, appearing as if closely arranged all round in alternating perfect and imperfect pairs as described by McMurrich (p. 40). In one example, where the disc was exceptionally transparent, thirty-six pairs of mesenteries reaching the stomodæum could be counted

The first part of a mesentery is narrow; it then thickens abruptly, the retractor muscle extending nearly across the face, again terminating in an abrupt manner in the imperfect pairs, but gradually in the perfect. The microscopic appearance of the retractor muscle is figured by McMurrich. All the mesenteries in section appear at first undulating on both sides, due to the enlargement of the mesoglœa, but become straight towards the stomodæum. On the face opposite the retractor muscle a thin musculature occurs all along, but the mesoglœa is not plaited. The parieto-basilar muscle appears developed on this face only. The endoderm is loaded with glandular cells, and fine nervous fibrillæ occur between the musculature and the denser peripheral parts of the endoderm.

The parietal mesenterial stomata are small, circular apertures, located a little distance from the column-wall just below the sphincter muscle; the perioral are somewhat larger.

Gonads were restricted in one specimen to the second cycle of mesenteries. Prof. MᶜMurrich (1889, p. 40) found that "the reproductive organs were present on all the mesenteries, with the exception probably of the directives."

This species is very abundant in the neighbourhood of the coral reefs around Jamaica, wherever these have been examined, sometimes partly buried in the sand, but more often attached to rocks. Isolated individuals may occur, but usually a number are closely aggregated, so close, at times, as to give rise to a polygonal outline of the discs, the result of mutual pressure. The associate habit and often the presence of more than two gonidial grooves are no doubt indicative of reproduction by fission; and Prof. MᶜMurrich obtained several specimens at the Bahamas in various stages of division. A small, brightly-coloured Crustacean has been found on one or two occasions living on the disc; but this commensalism is evidently not so constant a feature of the West Indian species as of those described and figured by Mr. Saville-Kent, from the Australian Barrier Reef. Here a brilliantly-coloured fish and one or two species of prawns are commensal with the polyps, and may pass in and out of the gastric cavity.

My reasons for regarding the *Discosoma anemone*, described by MᶜMurrich, as the *Actinia helianthus*, of Ellis, are given at the end of the description of the next species.

Genus.—HOMOSTICHANTHUS, n. g.

Discosomidæ, in which the tentacles are slightly knobbed and arranged in numerous peripheral cycles and radiating rows, a single row communicates with each endocœle and exocœle. Column-wall devoid of verrucæ; disc much folded. Two deep gonidial grooves. Sphincter muscle restricted.*

The generic term has reference to the practical similarity of all the rows of tentacles, both endocœlic and exocœlic.

*Prof. Haddon (1898, p. 432) employs this useful term for an endodermal sphincter muscle, in form intermediate between the diffuse and the truly circumscribed types. It refers to an intermediate stage, in which the mesoglœal plaitings, for the support of the musculature, do not arise from a common axis, but from several principal axes of less complexity. He further suggests "constricted" for the typical circumscribed muscle. (cf. figs. 3 and 7, Pl. xiii., "diffuse endodermal muscle"; fig. 6, Pl. xii., "restricted endodermal muscle"; figs. 1, Pl. xiv., fig. 2, Pl. xv., "circumscribed (constricted) endodermal muscle"; also, "aggregated," MᶜMurrich, 1893, p. 152, pl. xxii., fig. 23.)

Homostichanthus anemone (Ellis).

(Pl. x., fig. 8 ; Pl. xii., figs. 4–6 ; Pl. xiv., fig. 2 ; Pl. xv., fig. 1.)

Actinia anemone, . . Ellis, 1767, p. 436, pl. 19, figs. 4, 5 ; Ellis and Solander, 1786, p. 6, &c.

Hydra anemone, . . Gmelin, 1788, p. 3869.

Discosoma anemone, . Duchassaing, 1850, p. 9 ; Milne Edwards, 1857, p. 257 ; Duchassaing and Michelotti, 1860, p. 38, pl. vi., figs. 2, 3 ; Duchassaing and Michelotti, 1866, p. 122. Andres, 1883, p. 493. [non M'Murrich, 1889.]

The base is flat and usually buried in sand for some distance below the surface of the sea-floor ; or may be fixed to rocks, gravel, or other foreign bodies. It is thin-walled and semi-transparent, the radiating mesenterial attachments showing through. In diameter it is slightly larger than the lower part of the column, but much less than the upper overhanging region. Particles of sand and gravel may adhere, and occasionally remnants of a coarse cuticular membrane. When not attached, as in the laboratory, the base is very distensable, and preserved examples exhibit radiating and concentric basal foldings.

The column for its whole length is buried in sand, and is greatly elongated, somewhat cylindrical, erect, smooth, distensable, and devoid of verrucæ. Distally the internal attachment of the different orders of mesenteries is apparent through the thin wall ; an additional cycle of pairs is in this way seen to commence about half-way up the column, and to extend as far as the apex. The distal region is strongly folded, and, along with the disc, overhangs the proximal region ; *in situ* this area rests upon the sea-floor, or the whole polyp may be buried so that only the crests of the discal folds are visible. Around the apex of the column are small, obtuse elevations which may perhaps be regarded as acrorhagi. They are a little lighter in colour and correspond with alternate mesenterial spaces. A shallow fossa occurs between the circle of acrorhagi and the outermost cycle of tentacles. In preserved specimens the column is divided into deep longitudinal and trans-verse foldings. When alive the polyps are capable of considerable retraction, and, if disturbed, withdraw themselves for some distance below the surface of the sea-floor.

The disc is large, and peripherally is thrown into deep folds, nine to twelve, or even more, in number. The central naked area is comparatively small and flat, and the peristome but slightly raised. Generally the disc is only partly retracted so that its diameter is not larger than that of the column ; it can, however, be completely withdrawn so as to be wholly hidden.

The tentacles are short, smooth, slightly capitate, and arranged in numerous

2 B 2

nearly similar rows; in a good-sized specimen over 300 radiating rows were counted. The apex may be shortly and bluntly conical. All the tentacles are of about the same size, but may vary a little with the amount of distension, sometimes becoming quite flaccid on the withdrawal of water.

The tentaculate area of the disc is divisible into two very distinct regions—an outer, in which the tentacles form about twelve cycles, each containing the same number; and an inner, in which the rows begin to terminate at different distances from the mouth. Only thirty or forty rows are continued the full length of both regions, but all extend across the first. In large specimens no serial order is obvious in regard to the lengths of the inner rows, but three or four orders can be made out in young specimens (Pl. XII., fig. 4). Peripherally, the tentacles are so closely arranged that on a slight contraction of the polyp the apices press one against the other and assume a polygonal outline, and sometimes more than one row appears to communicate with a mesenterial chamber. Small developing examples may occur among the others, especially at the margin which appears to be a region of continuous active growth. Occasionally a bifurcated tentacle is come upon, and omissions may occur here and there, especially in the inner series. Only one row communicates with each mesenterial space. The organs possess considerable adhesive power when alive, though not so marked as in the former species, and can move about independently of one another.

The gonidial grooves are very distinct, the two enclosing lips being thick and protuberant. The walls of the stomodæum are slightly ridged, and so delicate that the mesenterial lines show through. In a state of repose the mouth is comparatively small and oval, but in preserved specimens it is widely open and nearly circular.

A small specimen, only about two centimetres in diameter, which I regard as an immature form, was found at Port Antonio adhering to a Thalassia leaf. It differed from the ordinary condition in having several tentacles distended to two or three times the size of the others, giving the disc quite a peculiar appearance.

The colour of the base may be faint scarlet, the intensity varying in different examples; or, it may be cream-white with only minute flecks of scarlet. In most the lower region of the column is a very bright scarlet or orange-red, sometimes in small patches on a cream ground. Distally the column is dark brown or steel grey. Occasionally the column-wall may be almost devoid of colour, except in the distal region, which always passes gradually into a very dark brown. The brown coloration is evidently determined by the presence of zooxanthellæ in the endoderm. Histologically it is shown that these occur only in the upper part of the column.

The disc varies much in colour in its different regions. Peripherally, where the tentacles are closely aggregated, it is usually a uniform light or dark yellowish-brown ; more centripetally, it is divided into narrow, radiating areas separated by dark lines. Each area consists of distinct, opaque white patches, or of continuous, opaque white bands, and corresponds with the rows of tentacles. The central, naked part of the disc is darker, and usually shows a purplish tinge, and a few white flecks may be scattered about. The margin of the lips is a stronger purple. The tentacles, both in different regions of the disc, and even in different parts of the same tentacle, also vary considerably. At their origin the short stems are of much the same colour as the portion of the disc from which they arise. Many of the inner show an opaque white circle at the place of origin. The tips of most are strongly coloured ; at their thickest part is an opaque white annulus, while the area immediately above may be greyish, yellowish-brown, or iridescent green. The last-mentioned condition is usually exhibited by the peripheral cycles, and the first cycle of this series often projects slightly beyond the others, its tentacles having intensely opaque white tips, which give a marked peculiarity to the colour-pattern of the disc. Usually the tips of the internal tentacles are whiter than those of the outer. At any part of the tentacular area, larger, bright green tentacles may occur.

The capitula of all the tentacles of several specimens obtained near the bathing place at Port Antonio were a bright-green, and the white opacity on the disc was absent, the whole surface, except the purple peristome, being a rich dark brown.

As mentioned by Duchassaing and Michelotti, the brighter colours are sometimes evanescent or may undergo modification. The rich, tentacular colours of some specimens kept in the laboratory, and exposed to the full sunlight for a few hours, practically disappeared, the whole disc and tentacles being reduced to a thin, opaque white and delicate brown. Others, especially those not so brightly coloured, showed no alteration.

The diameter of the expanded disc, in the living condition, varies from 10 to 15 cm., or may even expand to as much as 20 cm. The diameter of the column may be about 6 cm., but depends much upon the amount of distension ; the height is from 7 to 8 cm., but in a tall jar in the laboratory the column elongated to as much as 9 or 10 cm., and swayed to and fro. The length of the tentacles is 0·4 cm., and the greatest diameter, which is towards the tips, is 0·2 cm. The disc of specimens preserved in formol is about 5·5 cm. in diameter, and the column about 3·5 cm. in height, and the same in diameter.

Anatomy and Histology.

The basal ectoderm is a broad layer, constituted mostly of supporting cells, among which are a few granular gland cells. The mesoglœa is usually narrower than the ectoderm, and is finely fibrous in character, with

many included cells. On its inner border it gives rise to short, branching
processes for the support of the weak endodermal muscle. The endoderm is
much the narrowest of the three layers, and, in places, shows a distinct nervous
layer.

The column-wall is much and deeply folded, and is of only medium thickness,
the mesoglœa throughout being of about the same breadth as the ectoderm. The
fine ridges noticed among the external characters are seen in sections to be
produced either by coarse bulgings, or by long, narrow processes of the middle
layer. Long gland cells with granular, non-staining contents occur in the
ectoderm in much greater abundance than at the base. No ectodermal
musculature can be distinguished. The inner border of the mesoglœa is thrown
into very delicate, branching processes for the support of the endodermal circular
muscle; this extends the whole length of the column, and is often most strongly
developed in the lower part of the column. At the actual apex, however, it is
better developed, and gives rise to a feeble sphincter muscle of the restricted type.
The endoderm is fibrillar or reticular on its mesoglœal aspect; towards the free
border the cells bear zooxanthellæ, the nucleus of which stains deeply and is
highly refractive. From the lower stomodæal region downwards, the algæ are
practically absent from the columnar endoderm of the column, and none occur at
the base.

Considering the magnitude attained by the polyps, the sphincter muscle
(Pl. xii., fig. 6) is remarkably feeble. The fibres are arranged on a few, narrow,
branching mesoglœal processes, developed for some little distance along the apex
of the column-wall, the whole being intermediate in form between a circumscribed
muscle, such as that of Stoichactis, and a diffuse sphincter, as in Corynactis. In
truly radial sections, the middle mesoglœal processes are a little longer and more
branching than are represented in the partly tangential figure given, very closely
resembling those of *Radianthus macrodactylus* (H. & S.), figured by Haddon (1898,
pl. xxxi., figs. 2, 3). In complete retraction of the polyps, the disc can be entirely
hidden.

The tentacles are all alike in structure. The apex is crowded with long
narrow nematocysts showing the internal spiral thread; an occasional granular
gland cell also occurs. A marked histological difference is apparent between the
capitulum and the stem, nematocysts occurring only in the former. In the stem
the ectoderm is narrower, and gland cells are more numerous, while the mesoglœa
thins towards the apex. The endoderm is a thick layer, with irregular internal
boundaries; small zooxanthellæ are abundant, and less so glandular cells with
highly refractive contents. Both the endodermal and ectodermal muscles are
very feeble, and connected with the latter, a fibrillar and a nervous layer show
very distinctly at the capitulum. Nematocysts are absent from the ectoderm of

the disc, gland cells occur, and the endodermal muscle is much stronger than in the tentacles.

The stomodæum is greatly folded in all the sections, the ectoderm being followed by long, narrow or broad processes of mesoglœa. The first layer is densely crowded with large, granular gland cells extending completely across, but stinging cells are rare. A very delicate nerve layer can be discerned, and the merest trace of ectodermal and endodermal musculatures. Zooxanthellæ are scarce in the endoderm, but gland cells are numerous.

The gonidial grooves are interesting in the amount of histological detail indicated, and remarkable for the enormously exaggerated endoderm (Pl. xiv., fig. 2). The ordinary stomodæal ectoderm and mesoglœa narrow just before reaching the groove, and then all the three layers become much thickened, the endoderm most so. In the ectoderm, the nuclei are nearly all restricted to a narrow, extremely well-defined zone, a little below the ciliated margin; for a short distance within this zone the layer is almost clear, and then another nucleated zone is apparent, but in this case the nuclei are much fewer and do not stain so deeply. Then comes another clear zone, and afterwards a nervous layer from which fibrillæ extend to a very feeble muscle layer, apposed to the inner face of the mesoglœa. Ganglionic cells are scattered here and there among the fibrillæ. The whole succession of details can be easily traced all round the gonidial ectoderm. The mesoglœa is smooth on its ectodermal aspect, but the endodermal aspect is irregular; it is finely fibrous in structure, and many isolated cells are included.

The endoderm of the groove is enormously swollen, and of peculiar structure. Nearly all the nuclei and protoplasmic contents are aggregated towards its periphery, the greater portion of the layer appearing highly reticular in section; granular gland cells are scattered about, more numerous towards both its internal and external limitations. The mesoglœa of the directive mesenteries as it passes through the endoderm is extremely narrow.

Several specimens dissected transversely exhibit numerous pairs of mesenteries, arranged in four orders. The number is very variable, no two of the examples being alike. Two gonidial grooves and two pairs of directives were, however, present in each case. To the naked eye both sides of the groove are smooth, and readily distinguished from the rest of the stomodæum by being unfolded; the mesoglœa and endoderm are also much thickened. Twelve pairs of perfect mesenteries were present in a transverse dissection through the middle stomodæal region of a rather small polyp, and also second and third imperfect cycles. In places these exhibited the normal regularity, but in some of the exocœles additional imperfect pairs belonging to lower cycles occurred, and all stages in the development of new pairs could be traced. In another polyp between thirty and forty pairs of perfect mesenteries were counted in sections through the middle stomodæal region, while some of the free pairs were attached higher up. The members of the fourth cycle extended only a short distance from the column-wall. It would thus appear that the normal arrangement of the mesenteries in the stomodæal region of young polyps is as follows:—the first cycle of twelve pairs of perfect mesenteries constitutes the first and second orders; a second cycle is formed of twelve alternating pairs; and a third cycle of twenty-four pairs. Beyond this irregularities begin to step in. In older specimens many more than twelve pairs become united with the stomodæum. The region of the directives is always that of most forward growth.

The mesenteries present a concave outline as they cease their connexion with the stomodæum, so that in sections through the lower region of the latter the free edge of the mesenteries, bounded by a mesenterial filament, appears twice, one part being in connexion with the column-wall, and the other, shorter part with the stomodæum. The six mesenteries of the second series become free in advance of those of the first.

The retractor muscles extend across nearly the whole face of the mesentery, but are nowhere much thickened, resembling somewhat those of *S. helianthus*. They differ in this respect from those of *A. elegans*, which are circumscribed and project considerably.

The parieto-basilar muscle is strongly developed on both faces, and supported on numerous fine mesoglœal plaitings, but a separate pennon is not present, at any rate in the upper region. The nervous layer and fibrillæ are very distinct in this region.

The retractor muscle commences abruptly and extends along the greater part of the face of the mesentery, terminating more gradually centripetally. The mesoglœal processes are long, narrow, and branching, and constitute nearly the whole of the thickness of the mesentery. The endodermal epithelium is very narrow comparatively, and contains numerous granular gland cells. There is little trace of any oblique musculature. The mesenteries are very narrow beyond the retractor region.

The mesenterial filaments are typical in character, closely resembling those of *Phymanthus crucifer*, already described. The trilobed condition occurs on the first two or three cycles, and is continued for but a short distance below the aboral termination of the stomodæum. The Nesseldrüsenstreif or glandular streak at the apex of the middle lobe, is very limited in its extent, and the first portion of the intermediate streak is characterized by an abundance of small zooxanthellæ. The Flimmerstreifen or ciliated streaks also occupy but a small region of the lateral lobes. The cells of the reticular streak contain but little protoplasm, while the mesoglœal axis in all three lobes is crowded with small, deeply-staining cells.

Below the stomodæum the mesenteries branch considerably at their free extremity, each division being terminated by a simple, more or less rounded filament. As each side of the filament approaches the mesenterial epithelium, its cells stain more deeply, and, compared with the more apical region, fewer gland cells are discernible.

At various places around Port Antonio, on the north-east side of the island, the species occurs in some abundance, usually with the column and part of the disc buried in the sand or among the roots of various marine plants, such as Thalassia and Ruppia. In water of from 4 to 5 feet around the bathing place belonging to the Titchfield Hotel many specimens are to be found, including the green variety. It does not affect a social habit, as is often the case with the previous species, and is entirely absent from around the Port Royal Cays and other spots on the south side of the island.

Duchassaing and Michelotti met with the form at Guadaloupe and St. Thomas. Ellis records it merely from the West Indies.

Though there can be no doubt as to the distinctness of these two West Indian Discosomids, yet, owing to the incomplete descriptions and figures of the earlier authors, some difficulty exists as to their identification with one or the other of Ellis's species. Assuming that the distinctions between the two forms indicated by Ellis were simply due to different degrees of contraction, or to age, Professor M‘Murrich regarded them as synonymous, and describes both as *D. anemone.* Notwithstanding the few details given by the older writers, there is every likelihood that M‘Murrich's Bahaman representative is really the *helianthus* of Ellis, and also of Duchassaing and Michelotti, and this is the determination which I have followed above. Ellis mentions the flat salver-shape for *helianthus*, while Duchassaing and Michelotti refer to the greenish-brown verrucæ; again, Ellis records the angular form of disc of *anemone*, and the two later authors refer to the colour variation.

I follow Haddon (1898, p. 473) in transferring *helianthus* (= *anemone*, M‘Murrich) to his new genus Stoichactis, and find it necessary to erect another genus for *anemone.*

The following tabulated characters will enable the two to be readily distinguished in collecting:—

Stoichactis helianthus.

Polyps often closely associated. Column short, salver-like, verrucose, usually not embedded; little retractile power.

Disc flat; an outermost cycle of tentacles alternates with all the other rows. Occasionally more than two gonidial grooves.

Colour of tentacles mostly greenish yellow, with lighter and darker patches; undergo no rapid variation in intensity.

Homostichanthus anemone.

Polyps scattered. Column long, cylindrical, non-verrucose, usually completely buried; capable of considerable retraction.

Disc sinuous; a series of about twelve cycles of tentacles constitutes a distinct peripheral zone. Only two gonidial grooves.

Colour of tentacles bright emerald green, with opaque white and brown; stronger colours readily change in intensity.

Genus.—**ACTINOPORUS**, Duchassaing.

Actinoporus, Duchassaing, 1850; Duchassaing & Michelotti, 1860.

Discosomidæ, in which a radial tentaculate area, bearing more than one row of tentacles, communicates with each mesenteric chamber. Tentacles all vesicle-like, either simple or lobed, no distinction between a peripheral and an inner series. The column-wall is provided with verrucæ distally. Sphincter muscle strong and circumscribed. Mesenteries all complete. A weak ectodermal musculature on the column and stomodæum.

The genus was instituted by Dr. Duchassaing (1850, p. 10) for a single West Indian species of anemone, differing much in regard to its tentacles from any other known form. Later, in collaboration with Michelotti (1860, p. 46), he gives a further description of the genus, in which he evidently regards the tentacular areas as homologous with the frondose areas occurring in Oulactis, the internal cycles of ordinary tentacles, present in the latter, being wanting.

An acquaintance with these two genera demonstrates, however, that no such relationship can be sustained ; the frondose areas in Oulactis are of columnar origin, and occur outside the sphincter region, while those of Actinoporus are discal, and within the sphincter region.

The most salient character of the genus is the occurrence of more than one row of tentacles communicating with each endocœle and exocœle, a feature unique among the Actiniaria, unless the same may be said of Actinodendron and of *Discosoma ambonensis.* Kwietniewski (1898, p. 410) describes the tentacles of the latter as in radial groups, a condition which seems to me, should certainly warrant at least the generic separation of the form from other Discosomæ, in which only one radial row communicates with each mesenterial chamber.

In the tentacular areas of the oral disc one may perhaps see some relation of degree between this genus and Actinodendron. In this latter, as figured and described by Haddon (1898), the forty-eight tentaculate areas, which likewise correspond with both the endocœles and exocœles, are prolonged for some distance as non-retractile lobes, and the tentacles on them are small, arise in an irregular manner, and are dendritic or form " conical bossy agglomerations." One may perhaps regard the lobes of Actinodendron as extensions of the sharply defined areas in Actinoporus, and the dendritic tentacles as exaggerations of the vesicular outgrowths in the West Indian genus.

From an acquaintance with only a single specimen of Actinoporus, it would be premature to regard the possession of only one gonidial groove (monoglyphic), associated with two pairs of directives, as a constant generic character.

Actinoporus elegans, Duchassaing.

(Pl. x., fig. 9 ; Pl. xɪ., fig. 8 ; Pl. xɪɪɪ., figs. 2, 6 ; Pl. xɪv., fig. 3 ; Pl. xv., fig. 2.)

Actinoporus elegans,	Duchassaing, 1850, p. 10; Milne-Edwards, 1857,
	p. 277 ; Duchassaing and Michelotti, 1860,
	p. 46, pl. vii., fig. 6 ; 1866, p. 132.
Aureliania elegans, . . .	Andres, 1883, p. 497.

The base is buried to a considerable depth in sand and gravel, and is thin-walled, the lines of attachment of the mesenteries showing through ; towards the margin it may also be deeply grooved in correspondence with the mesenteries. In diameter it is scarcely larger than the column.

The column is greatly elongated, cylindrical, smooth, strongly ridged and grooved above and below, and, but for a thin opaque whiteness, nearly transparent. A row of circular transparent verrucæ occurs on all the ridges, rendered very evident by the absence of the opaque whiteness, they appear more like vesicles in the preserved polyp. Along some ridges the transparent discs are not so perfectly circular as on others, and they may be in more than a single series, or even become contiguous. In the preserved condition the column is coarsely wrinkled transversely, less so longitudinally, and is of greater diameter above than below. A smooth, deep fossa exists between the marginal verrucæ and the tentacles.

The disc is flat or partly folded, not much broader than the column, and made up of forty-eight long, radiating, triangular areas, separated one from the other by deep, smooth sulci. A small, central area is naked and smooth. The areas bear extremely short, capitate or spheroidal tentacles, which seem to be little more than small vesicular outgrowths of the disc. These are often bifurcated or lobed, and extend from the fossa to near the mouth, increasing a little in size from within outwards. Odd smaller vesicles occur among the larger. Over the greater part of the disc the tentacles arrange themselves approximately in two rows along each side of a radiating area, but they communicate in an irregular manner with the cœlenteron. Towards the centre of the disc they form but a single row along the middle of the radiations. Some rows extend slightly more centrally than others, but no serial arrangement can be distinguished.

As a whole the tentacles give to the disc, both in the living and preserved condition, a finely beaded appearance ; peripherally they completely hide its actual surface, but are more distant towards the middle. They possess apparently no power of retraction, and communicate with both the endocœles and exocœles,

176 J. E. Duerden—*Jamaican Actiniaria :*

though a slight disparity occurs in that forty-eight rows occur, while there are twenty-five pairs of mesenteries, forming, of course, fifty mesenterial chambers. The disc overhangs the column a little, and can be almost completely retracted.

The base presents a very delicate opaque whiteness, and nearly the whole superficial area of the column shows a similar opacity, particularly evident in the upper region; clear and transparent areas may, however, remain in places, as at the verrucæ. In the more distal region some of the ridges may be a delicate transparent brown. The colours of the knobs of the tentacles are variable, and not arranged according to any definite pattern. They are mostly opaque white, with various mottled colours on a clear transparent ground; spots of yellow, brown, pink, red, and white are irregularly mingled. The marginal tentacles are more spotted with opaque white than are those more internal. Preserved in formol the specimen changed its colour as a whole to a dark brown.

The column may extend to as much as 15 cm. (6 inches) in height, and is about 5 cm. in diameter.

The ectoderm of the base is an exceptionally deep layer. Large numbers of long, unicellular glands occur of about the same diameter throughout, and contain finely granular matter.

The mesoglœa is narrower than the outer layer, and is slightly fibrous in character; numbers of small cells are included within it. Internally it is finely plaited for the support of the endodermal muscle, which is here feebly developed. The endoderm is the narrowest of the three layers, and presents irregular internal limitations, and many granular gland cells.

The column-wall is much and deeply folded, and of moderate thickness in each of its three layers. Peripherally the ectoderm appears somewhat dense, owing to the great abundance of unicellular glands with finely granular contents ; the cells extend from the inner limits of the layer, but become more swollen towards the outer surface. Though the polyp was nearly transparent when alive, the column-wall changed to an opaque dark brown on preservation in formalin, and the contents of the ectodermal cells appear yellowish brown on microscopic examination. The nuclei of the ectodermal supporting cells are mostly aggregated within a middle zone, and a slight ectodermal musculature is developed.

The fibrous nature of the mesoglœa is more obvious in the column than at the base, and the layer is very irregular in its outline, giving rise to numerous, deep folds on both its outer and inner aspects. On its endodermal border it forms, in addition, long, narrow, branching processes for the support of the strong, endodermal, circular muscle. Towards the proximal region of the column, these processes become more numerous and longer, being even longer than the mesogloea is broad.

The endoderm is devoid of zooxanthellæ, and such is the case throughout the polyp. For some distance from the mesogloea it is constituted largely of delicate fibrils, while the protoplasmic contents are aggregated near the free, irregular border. A nervous layer is distinctly shown in places.

Sections through the verrucæ present no histological difference from the rest of the column-wall, except that all three layers are thinner and the musculature is weaker. It is doubtful as to how far they can be compared with the verrucæ in the species already described. They certainly possess no adhesive power.

The ectoderm at the fossa is strongly ciliated, the cilia being still obvious in preserved material, though this is not the case elsewhere on the column.

The sphincter muscle (Pl. xv., fig., 2) is an enormous, circumscribed, endodermal representative, hanging by a very narrow base from the floor of the fossa. Even to the naked eye it is a very pronounced outgrowth, 7 mm. in length. It breaks up into many large lobes, the appearance differing much in different sections, and in places seems to enclose portions of the coelenteron. A narrow mesogloeal axis extends down the middle, and from it branching processes arise in a somewhat pinnate manner, and are continued into each lobe. A peduncle is practically absent, and the mesogloea of the sphincter is in continuity with that of the column-wall only within very narrow limits, the ordinary endodermal muscle being traceable nearly across the connexion. The mesogloeal processes branch very much; the lining muscle fibres are not represented in the figure. Though differing in detail, and many times larger, it will be seen, on comparison of the two figures, that the muscle is exactly of the same type as that in *S. helianthus* (Pl. xiv., fig. 1). It is probably the largest circumscribed sphincter known in any Actinian.

In radial sections through the disc, the tentacles are displayed as crowded, irregular, thin-walled, vesicular outgrowths of the disc; and, compared with those of the column, each of the three component layers undergoes some modification. The ectoderm loses its gland cells, and the outer half is constituted almost entirely of a clearly-defined zone of small nematocysts; the inner half of the ectoderm, on the other hand, appears as a nuclear zone. Neither an ectodermal nor an endodermal musculature is distinguishable, and the mesogloea is nowhere plaited. The endoderm is very narrow, and brown pigment granules take the place of zooxanthellæ.

Where a section passes through a group of tentacles (Pl. xiv., fig. 3), the disc is indistinguishable from the tentacles themselves, and the two are practically alike in structure, the disc being thin-walled and possessed of a nematocyst layer; a feeble circular endodermal muscle is, however, present in the perioral region of the disc.

The tentacles in this species therefore differ from those of the two previous forms in not having a capitulum histologically distinct from a stem, and also in not being much differentiated in structure from the disc itself.

Towards the middle of the disc, that is, in the naked area, the details, however, approach more closely those of the column; gland cells, with highly refractive contents, occur in the ectoderm, and the endodermal muscle is stronger.

The wall of the stomodæum is much folded, both vertically and transversely. The ectoderm is broad, ciliated throughout, and bears numerous, long, granular, gland cells, and a less number of oval-shaped nematocysts, much larger than those of the tentacles. A very weak ectodermal musculature is discernible. The endoderm is slightly pigmented like that of the disc and tentacles, and contains a few highly refractive gland cells.

A transverse section through the polyp, in the middle stomodæal region, shows to the naked eye the following details :—

Twenty-five pairs of mesenteries, all of which are complete; of these two pairs are directives, so that of the other pairs, twelve occur on one side, and eleven on the other. No incomplete mesenteries are developed. A single deep gonidial groove, with very smooth walls and much thickened mesoglœa, is included between one of the pairs of directives, but no indication of a second is presented in connexion with the opposite pair.

The inner mesenterial stomata occur just within the lips, and the outer a little from the column-wall, about a centimetre below the sphincter muscle. Both are rather large apertures of about equal size. The retractor muscles of the mesenteries are large, thick, oval, or reniform projections from one face, and are attached by only a narrow, short pedicle; on the opposite face of the mesentery a very distinct pennon arises a little beyond the insertion of the mesentery in the column-wall.

In a section below the stomodæal region, the same twenty-five pairs of mesenteries occur, and, with their mesenterial filament and gonads, completely fill the cœlenteron. Towards the basal part of the column, alternate larger and smaller pairs are exhibited, but irregularities occur, in one region seven pairs being of the same size. The number of mesenteries bearing mesenterial filaments begins to diminish, until a little above the base they are met with only on four. In the single example studied, the gonads were bright red in colour, and a reddish oil was extracted by alcohol.

The microscopic appearance of a portion of a mesentery, near its place of origin in the column-wall, is represented in Pl. xiii., fig. 2. The pennon is seen to be strongly developed, and the mesoglœa long and deeply plaited on both sides. In the upper region of the column, from which the figure was taken, the pennon is near the wall, but below it becomes further and further removed as the parieto-basilar muscle becomes stronger.

The enormous circumscribed retractor muscle of each mesentery is arranged on branching, mesoglœal processes, the mesoglœal axis from which they arise being very thin. The muscle layer is continued along the face of the mesentery beyond the swelling as far as its connexion with the stomodæal wall, but the mesentery, as a whole, is very thin, both before and beyond the enlargement. The mesenterial endoderm contains abundant, deeply-staining, granular cells, and a nervous layer is distinctly separable in places, especially near the pennon.

The mesenterial filaments are trilobed in the upper region, and exhibit the usual details of structure. The glandular and intermediate streaks are densely crowded with gland cells with brown granular contents. Proximally the middle lobe becomes highly glandular; and the mesenterial endoderm immediately behind is swollen. Its cells, along with those of the mesenterial epithelium, contain much dark granular matter.

Female gonads occurred on prolongations of some of the mesenteries, but, owing to the crowded condition of the cavity it was impossible to determine their precise arrangement. In the gonad region the mesoglœa of the mesenteries becomes extremely thin, the endodermal epithelium is much broadened, and the contents of the cells highly granular in character, while pigment granules and granular gland cells occur along the margin.

So far as could be determined from dissections, the twelve pairs of mesenteries constituting the first and second orders, and including the directives, are fertile.

I have identified this peculiar species as the *Actinoporus elegans,* of Duchassaing, although certain differences call for notice. The colour in the Guadaloupe specimens is stated to be blue, and the tentacles reddish white, while the length is given as 35 mm. It is a species which suggests the possibility of much colour variation, but it seems a little remarkable that the Jamaican specimen should be three or four times larger than the others.

Only a single specimen was obtained from along the shore to the east of Wood Island, Port Antonio, during the temporary establishment at the latter place of a Marine Laboratory in connection with the Johns Hopkins University. This was collected by Dr. H. L. Clark, and kindly handed over to me. Although the locality was afterwards carefully searched on many occasions, no other example could be found. The column was buried for a considerable distance in the muddy sand, the disc alone being exposed. Large specimens of Asteractis, which have the same habit, occur in the vicinity. The skeleton of a crab, with all the flesh digested, was extruded by the animal while under observation in the laboratory.

MM. Duchassaing and Michelotti obtained their specimens upon submerged rocks at Guadaloupe.

Andres places the species, as *Aureliania elegans*, among his *Aurelianidæ dubiæ*. It is undoubtedly a Discosomid as here defined, and not enough is yet known of the British Aureliania to warrant such a generic relationship. The perfect similarity of type of its sphincter muscle with that of *S. helianthus* must be taken into account in any consideration of its relationships.

Family.—CORALLIMORPHIDÆ, Hertwig.

Corallimorphidæ,   Hertwig, 1882 ; McMurrich, 1893 ; Haddon, 1898.
Corynactidæ,   .   Andres, 1883.

Stichodactylinæ, in which the tentacles are all of one form, capitate, and comparatively few ; a distinction between a peripheral cyclic series and an inner radial series may or may not be apparent. Muscular system weak in all parts of the body ; sphincter muscle absent or weak.

This family was established by Professor R. Hertwig (1882, p. 21) for the reception of two species of "Challenger" Actiniæ, both belonging to the genus Corallimorphus of Moseley (1877). The genus was considered to bear a close relation to both Discosoma and Corynactis, and, in the "Supplement" (1888, p. 10), the latter is definitely included in the family. In his great work, published a year later, Andres employed the more preferable family name Corynactidæ to embrace the genera Corynactis, Corallimorphus, and Capnea. Of the two terms having thus practically the same significance, Hertwig's, bearing priority, must be the one employed.

The characters which Hertwig regarded as of greatest diagnostic importance in the genus Corallimorphus, and which at that time held also for the family, were, "the double corona of tentacles, the equal distribution of the reproductive elements, and the absence of the circular muscle." These can now be retained only for Corallimorphus. In Corynactis there is not the same distinction between an outer and an inner series of tentacles, the distribution of the gonads is not fully known, while a circular muscle, though not strong, certainly exists.

Genus.—**CORYNACTIS**, Allman.

Corynactis,   .   Allman, 1846; Johnson, 1847; Milne-Edwards, 1857 ; Gosse, 1860; Andres, 1883; Hertwig, 1888 ; Haddon, 1898.
Draytonia,   Duchassaing and Michelotti, 1866.

Corallimorphidæ, in which the column-wall is smooth, tentacles knobbed, arranged in several cycles and in radiating rows, the outer larger than the inner. Tentacles and mesenteries generally tetramerous. Gonidial grooves present or absent. An ectodermal longitudinal muscular layer on the column-wall and on the stomodæal wall ; endodermal sphincter muscle very weak. Mesenterial filaments devoid of ciliated streak. Mesoglœa practically homogeneous.

All the representatives of the genus are small polyps, and the descriptions of the various species given by the older writers refer only to the external features. The most salient of these are the knobbed tentacles, and the communication of more than one with a mesenterial space. In the present species, and in *C. viridis*, they are tetramerous, but in *C. carnea*, Stud., according to Kwietniewski's (1898) observations, the mesenteries appear to be hexamerous.

Professor Hertwig (1888) was the first to make a histological examination of any member of the genus, and to discover the presence of longitudinal muscle fibres on the outer side of the body-wall. Professor Haddon and myself (1896, p. 152) found the same in *C. australis*, and Haddon (1898, p. 467) in *C. hoplites*.

MM. Duchassaing and Michelotti erected the genus Draytonia for the species about to be described, distinguishing it from Corynactis on account of the circle of green spots upon the capitulum and disc. In the specimens which I have examined, the pigment spots do not project above the smooth surface, and those on the column cannot be in any way regarded as acrorhagi, and may or may not be present. They are certainly not deserving of generic distinction.

Haddon (1898, p. 468) refers to the curious fact that an "emerald green ring round the capitulum is characteristic of forms so widely distributed as European Seas (*C. viridis*), off the coast of Buenos Ayres (*C. carnea*), and Port Phillip, Australia (*C. australis*)." In the present instance the ring is represented by a circle of spots of the same colour.

Corynactis myrcia (Duchassaing and Michelotti).

(Pl. x., fig. 10; Pl. xii., fig. 7; Pl. xiii., figs. 3-5 ; Pl. xv., fig. 3.)

Draytonia myrcia,   .   Duchassaing and Michelotti, 1866, p. 124, pl ii., fig. 8.
Corynactis myrcia,   .   Andres, 1883, p. 485.

The base is spreading, and in diameter larger than the column. It is irregular in outline, tough, and firmly adherent to various objects. Polyps are met with in groups, and are occasionally found connected, one with another, by a thin basal expansion or cœnosarc. The column is short, mammiform or cylindrical in retraction, irregular in outline below, and circular or oval above. The surface is smooth, and the walls thin and translucent, the lines of attachment of the mesenteries showing through, and dividing the whole column into slight ridges and furrows.

The tentacles are arranged in cycles and in radiating rows, each row communicating with one mesenterial space. In one specimen in which the tentacles could be counted they numbered 48, arranged in four cycles in the formula 8, 8, 16, 16. The innermost tentacles are very short, appearing as mere tubercles on the disc; the intermediate show a distinct stem and knob, while the outermost are still larger, and overhang in extension; the stems are conical, and the knobs rounded. The disc is oval or circular, smooth, thin-walled, and nearly transparent, the mesenterial lines showing through; the oral cone may be very prominent. The mouth is oval; the walls of the stomodæum are deeply ridged and furrowed, and very protrusible; no gonidial grooves are indicated. The disc and tentacles may be entirely hidden in retraction.

The column is brown below, and almost colourless or crimson above; a circle of small, emerald green, capitular spots may or may not be present. The stems of the tentacles are translucent and colourless or yellowish; the knobs rose, red, or orange. The disc is brown, with white radiating lines; the peristome bears a narrow circle of emerald green spots; the stomodæal wall is white.

When retracted, the polyps measure about 0·7 cm. in diameter, and are the same in height.

Anatomy and Histology.

Examined histologically the ectoderm of the base and also of the column-wall is remarkable for the abundance of large unicellular mucous glands, mingled with the narrow supporting cells. They appear to constitute the greater part of the layer, becoming more swollen towards the free surface, where they give rise to an almost clear zone. The contents are sometimes clear and homogeneous, and do not stain in borax carmine; in most cases, however, they are finely granular, and take up the colour slightly. The nuclei of the supporting cells are arranged in a zone a little within the middle of the layer, while the most internal region of the ectoderm exhibits nerve and muscle fibrils. The ectodermal musculature is discernible on the base, but becomes stronger on the column-wall, the cut ends of its fibrils appearing as a very distinct layer in transverse sections.

Throughout the base and column-wall the ectoderm remains a high columnar epithelium, and in the latter is considerably folded in preserved material, the foldings being followed by the mesoglœa. A cuticle is not distinguishable.

The mesoglœa is very variable in thickness, owing to its numerous foldings. It is remarkably clear and homogeneous; no fibrillar structure is indicated, and it is practically devoid of any structural elements, isolated cells occuring with extreme rarity. Even after the other structures of the polyps have been deeply stained the mesoglœa remains colourless and indistinguishable from the field of the microscope. The same phenomenon is presented throughout the polyp, and evidently throughout the genus, and is that characteristic of the mesoglœa of the Madreporaria.

The endoderm of the base and column is scarcely narrower than the ectoderm, and gland cells with dense, highly-refractive contents, occur sparingly. Zooxanthellæ are absent throughout the polyp. The circular endodermal muscle is feebly developed in the base, but becomes stronger in the column, being supported on small mesoglœal plaitings, and enlarges distally to form the sphincter muscle.

The sphincter muscle (Pl. XIII., fig. 3) is endodermal, and intermediate in character between a diffuse and restricted form. The muscles fibres are very strong and closely arranged, and become concentrated on long mesoglœal processes; these latter, however, never become so long and branching as in *C. viridis* and *C. carnea*, but are stronger than those of *C. australis*.

The knobs of the tentacles consist almost wholly of a very deep ectoderm; the mesoglœa and endoderm extend into them but a short distance, and they never exhibit any lumen.

An outer broad zone of the ectoderm is largely made up of very long, narrow nematocysts showing the internal spiral thread distinctly. Occasionally a large, oval, stinging cyst is also seen, and in the deeper parts of the ectoderm are abundant, oval-shaped, deeply-staining bodies, evidently nematocysts in various stages of development, though some are granular gland cells. The endoderm contains an extraordinary quantity of granular pigment matter.

The ectoderm of the stems (Pl. XIII., fig. 5) is devoid of nematocysts, and in structure is much like that of the column-wall, being highly glandular. A weak ectodermal musculature is supported on branching, mesoglœal processes. From the muscle fibres on these processes very delicate fibrils radiate in a peculiar brush-like manner. The muscle is better developed proximally than distally, and is practically absent from the knob. Internally the mesoglœa forms deep, rounded plaits recognizable in longitudinal sections. The endodermal muscle is very weak, and the endoderm contains but little of the granular matter so abundant in the knob: the tissue nearly fills the lumen in retracted examples.

The disc closely resembles the stem of the tentacles in structure; the mesoglœa is delicately plaited for additional support to the endodermal and ectodermal musculature.

The stomodæum is oval in transverse sections, and the ectoderm is thrown into about twenty very deep and regular longitudinal folds, the mesoglœa also following. In position the folds bear a rough approximation to the points of attachment of the complete mesenteries. No indication of gonidial grooves is presented. Haddon (1898, p. 468), on the other hand, found three gonidial grooves in one example of *C. hoplites*, and one in another.

The stomodæal ectoderm is uniformly ciliated, and the supporting cells give rise to the usual zone of brightly-staining nuclei; several varieties of nematocysts are represented, and various kinds of elongated, granular gland cells.

Following the folds of the ectoderm the mesoglœa is very thick and triangular in transverse section, but between the folds it becomes extremely narrow; a weak musculature occurs on both its ectodermal and endodermal surfaces. The endoderm is a broad layer, constituted largely of gland cells, some with clear contents, and others which are granular and stain readily.

The mesenteries are tetramerous and arranged in three cycles; eight perfect pairs, of which two pairs are directives, represent the first and second cycles, and are about equally developed, while eight, incomplete, alternating pairs represent the third. In the upper region of the stomodæum an odd member of the free series may be connected with the stomodæal-wall for some distance, and, in one or two cases, remains attached for practically the whole stomodæal extent. This is especially noticeable in the region of one of the pairs of directives, as compared with the lateral pairs. In one instance an odd member of two pairs of the third cycle is connected with the stomodæum throughout its length, but on one side it is the mesentery next the directives, while in the other it is the next but one which is perfect. This latter condition is shown in Pl. XIII., fig. 4, where the second mesentery from the left, belonging to the third order, has just ceased its connexion at the termination of the stomodæum. Otherwise the regularity of the mesenteries in this species is in striking contrast with the lack of symmetry met with in other representatives of the genus.

The mesoglœa of the mesenteries is thick for some distance from its origin in the column-wall, and on one side it then forms plaits of greater or less complication (Pl. xv., fig. 3). These are of the same character as in *Corynactis australis*. The folds support the vertical retractor muscle, which also extends along the whole face of the mesentery. In no case does the muscle give rise to a thickened band, as in most members of the Actiniæ.

Beyond the region of greatest plaiting the mesoglœa narrows considerably; and here, in the proximal part of the polyps, the surface of the mesoglœa of the opposite face also becomes delicately plaited for the support of the oblique musculature. Distally the mesenterial endoderm closely resembles that of the column-wall, but contains a greater number of the glandular cells, with highly refractive contents. Proximally it becomes swollen, and contains many granular particles of various sizes, while elongated gland-cells are still more numerous.

The parieto-basilar muscles are developed for some distance along each face of the mesentery, and are continued a short way on to the column-wall. There is no trace of any mesoglœal folding or pennon.

Both cycles of perfect mesenteries remain connected as far as the inner termination of the stomodæal-wall. I have not been able to determine any definite order in which they become free, but the directives at one end remain united further than the opposite pair, the laterals being the first to cease their connexion.

The imperfect mesenteries project for some distance within the cœlenteron. In section they are almost as broad as the complete mesenteries, and the mesoglœa terminates in numerous processes, each surrounded by a muscular layer, which, so far as it extends, is as strongly developed as on the first cycle. All the endocœlic and exocœlic spaces are practically equal, and the mesenteries by no means fill the cœlenteric space.

The terminal edge of the stomodæum is reflected as a whole, so that in sections through this region the wall is double in all its three layers (Pl. xiii., fig. 4). The reflected ectoderm passes for some little distance outwardly along each face of the mesenteries, and appears in perfect continuity with the tissues forming the mesenterial filaments. At first the filaments are very irregular and narrow in outline, forming only a slightly rounded termination to the mesenteries; they are, however, histologically very distinct from the rest of the mesenterial epithelium. Lower they become more characteristic, and are either rounded or cordate in section (Pl. xv., fig. 3).

The mesenterial filaments are remarkable in that only the middle terminal lobe, the glandular streak or Nesseldrüsenstreif, is ever developed; the lateral lobes, bearing the ciliated streak or Flimmerstreifen, so characteristic of most Actiniaria, are never produced.

A little below the aboral termination of the stomodæum the mesenterial endoderm is considerably swollen immediately behind the filament, so as to produce in section somewhat the appearance of a trilobed filament; but these enlargements cannot be regarded as at all comparable with the lateral lobes of the more usual Actinian filament. Except in length of the constituent cells they differ in no important respect from the remainder of the mesenterial epithelium into which they often graduate insensibly, while the distinctly lobed character is not presented by all the mesenteries. Further, the mesoglœal axis never sends a branch into these swellings for the support of the cells, as is the case where the Flimmerstreifen are developed. Such a filament is characteristic of the Madreporaria, and occurs also in the proximal region of many Actiniaria after the ciliated streak has ceased to exist.

The mesoglœal axis of the mesentery, completely surrounded by a weak musculature, passes into the base of the filaments, and there becomes slightly expanded, sending a branch to each side. The cells on the anterior or inner border are much longer than on the sides and behind. They are mostly strongly ciliated supporting cells, with which are mingled glandular cells of various kinds, and large oval nematocysts with a loose internal thread. The latter are of the same form as occur more rarely in the ectoderm of the stomodæum and knobs of the tentacles.

In the stomodæal region the imperfect mesenteries are devoid of filaments; they appear, however, immediately below and completely resemble those on the chief mesenteries.

Proximally the mesenteries branch at their free termination, each branch being capped by a filament in which the large nematocysts predominate.

No gonads were present in any of the polyps examined.

On one occasion six specimens were collected at Drunkenman Cay, all closely associated within a crevice in the coral rock in shallow water; and another time several polyps were come upon living together on a live Pinna shell from Harbour Head, Kingston Harbour. When irritated they are capable of sending out quantities of clear mucus.

The species was first obtained by Duchassaing and Michelotti from St. Thomas, and described by them under the term *Draytonia myrcia*; Andres places it among his "Corynactidæ dubiæ," under the genus Corynactis. The correctness of this generic transference I have already referred to.

Its histological characters should be compared with those of *C. australis* (1896, pp. 152–3), and it will be seen that the two closely agree. The mesenteries are, however, more regular, and the sphincter muscle slightly better developed in the present species. The sphincter also differs from that of *C. viridis*, Allm. (1896, pl. viii., fig. 11).

The connexion of one polyp with another by a basal expansion, and the usual occurrence in groups are indicative of asexual reproduction, a method already known to occur in the British Globehorn, *C. viridis* (1860, p. 291). The irregularities in the arrangement of the mesenteries noted in *C. australis* (1896, p. 152), and in *C. hoplites* (1898, p. 468), are probably also due to this process. In the Australian representative it was found that some specimens possessed only one pair of directives, while others had two.

Attention should be directed to the tetrameral arrangement of the mesenteries, corresponding with the tetrameral tentacles; the extraordinary development of gland cells, both in the ectoderm and endoderm is noteworthy; and also the very large, oval cnidocysts in the knobbed tentacles, stomodæal ectoderm, and mesenterial filaments. The mesoglœa is an exceptionally homogeneous layer, and the retractor muscle of the mesenteries is arranged on only slight mesoglœal folds, never becoming circumscribed. Undoubtedly one of the most important anatomical features is the absence from the mesenterial filaments of any lateral lobes bearing the Flimmerstreif.

Tribe.—ZOANTHEÆ, R. Hertwig, 1882.

Family.—ZOANTHIDÆ, Dana, 1846.

Sub-family.—MACROCNEMINÆ, Haddon and Shackleton, 1891.

For the definitions of the Tribe, Family, and Sub-family, the first instalment of this series should be consulted, or better, the original papers of Haddon and Shackleton (1891, 1891 a).

Genus.—**PARAZOANTHUS**, Haddon and Shackleton, 1891.

Macrocnemic Zoantheæ, with a diffuse endodermal sphincter muscle. The body-wall is incrusted. The ectoderm is continuous. Encircling sinus as well as ectodermal canals, lacunæ, and cell-islets in the mesoglœa. Diœcious. Polyps connected by thin cœnenchyme, rarely distinct.

The characters of greatest generic importance are the macrocnemic arrangement of the mesenteries, a feature shared with the genus Epizoanthus, and the presence of a diffuse endodermal sphincter muscle.

In their "Review of the British Actiniæ," Haddon and Shackleton assign to the genus three European forms; *P. axinellæ* (Schmidt), *P. anguicomus* (Norm.), and *P. Dixoni*, n. sp., and, in their Report on the Zoantheæ collected by Professor Haddon in Torres Straits, make an addition of two new species, *P. dichroicus* and *P. Douglasi*. In the paper on Jamaican Zoantheæ, I show that the *Gemmaria Swiftii*, of Duchassaing and Michelotti (1860), must be transferred to Parazoanthus, and also advert to the fact that Carlgren (1895) has demonstrated that the supposed Antipatharian genus Gerardia, Lac.-Duth, must probably be regarded as belonging to the same genus. Reviewing the Zoanthean genera in his latest paper, Haddon (1898, p. 408) confirms Carlgren's statements with respect to this form, and locates Gerardia between the genera Parazoanthus and Epizoanthus.

Recent trawling in the Caribbean Sea has brought up from the Pedro Banks distant about 50 miles south-west of Jamaica, a branching Hydroid over 100 cm. in height, the trunk and main divisions of which are entirely incrusted with a single Zoanthid colony. It bears a very close external resemblance to *Parazoanthus dichroicus*, Hadd. and Shackl., but histological characters reveal that the two are quite distinct. I propose to term it *Parazoanthus tunicans*, on account of its investing habit.

Quantities of sponges were also trawled on the same occasion, many of which displayed small commensal anemones distributed over nearly their whole surface. On some massive, black sponges, two or three feet in diameter, the polyps appeared as distinct, white, circular discs, but on a dark, purplish sponge they were in small colonies, producing short catenulations.

A detailed study discloses that these two, though somewhat similar in their habit, are distinct species.

Anatomical examination leaves no doubt that the first sponge-incrusting form is a Parazoanthus, and I propose to term it *P. separatus*, in emphasis of the distinct character of its individual polyps. With regard to the generic position of the second, some uncertainty prevails. Owing to the remarkable shortness (0·5 mm.) of the polyps, and the presence of numerous large sponge spicules in the capitular region, I have failed to make out the arrangement of the mesenteries, or to discover any sphincter muscle.

Considering, however, the extreme weakness of the musculature in all the other parts of the polyp, and the thinness of the mesoglœa in the capitulum, there can be no doubt that any sphincter occurring will conform to the type characteristic of Parazoanthus; and further, comparing all its external and anatomical features with those already known in other species, I have little or no hesitation in assigning the form to the present genus. I propose for it the term *Parazoanthus monostichus*, the polyps being usually arranged in a single row.

A comparative study of the different representatives of the genus calls for a few remarks of more general interest in Actinian morphology.

In respect to both its musculature and the mesenterial filaments, Parazoanthus displays conditions which lead one to place it as the lowest of the Zoanthean genera, a position already assigned it by Haddon and Shackleton on less conclusive grounds. Taking the mesenteries only into account Haddon considers the other sub-family—the Brachycneminæ—may be regarded as slightly more primitive.

The musculature in all the species of Parazoanthus is weak. This is especially true of the sphincter muscle. In all other genera of the Zoanthidæ the sphincter is embedded in the mesoglœa, and is usually of considerable strength; in Zoanthus it is even double, being subdivided into an upper and a lower portion.

The diffuse, endodermal sphincter characteristic of the genus represents merely a concentration in the capitular region of the circular endodermal muscle which lines the column, usually throughout its length.

As the sphincter becomes more strongly developed, the mesoglœal foldings supporting it become deeper and deeper to give increased area for its support, until they may ultimately unite at their free edges, the muscle thus passing from the endodermal to the mesoglœal stage.

In *P. Swiftii*, where the general endodermal musculature and the sphincter are comparatively well developed, small portions of the latter do actually appear to become cut off from the endoderm and become wholly included within the mesoglœa (1898, pl. xx., fig. 5). The mesoglœal bays are, however, so deep as to suggest the possibility that the appearance of included muscle fibres may be merely a result of the direction in which the section is taken. Such would be the case if the depressions were deep and oblique to the plane of the section. The endodermal muscle is, of course, mesoglœal at the origin of a mesentery in the column-wall.

Concerning the sphincter of *P. dichroicus*, Haddon and Shackleton (1891 *a*, p. 699) remark:—"Near the upper extremity (in contracted specimens), it appears to become embedded in the mesoglœa, a few simple cavities being visible in our sections." We thus possess in the genus, indications, at any rate, of how the actual transference from an endodermal to a mesoglœal muscle is effected (cf. Haddon, 1898, p. 432).

The musculature is everywhere very feeble in *P. separatus*, and the sphincter certainly remains entirely endodermal. The polyps of *P. monostichus* are only about half the size of the former; and such a muscular weakness is indicated in all the organs that, independently of the interference of the incrustations, the sphincter would probably be difficult of recognition.

Two of the four Antillean species of Parazoanthus which I have examined, exhibit in their mesenterial filaments a simpler condition than that characteristic of other Zoanthids.

The structure of the upper part of the Zoanthean filament is well known. It is trifid or V-shaped in transverse section. A middle apical portion of ciliated supporting cells, granular gland cells, and nematocysts, constitutes the glandular streak, Drüsenstreif, or Nesseldrüsenstreif; the outer layer of the two lateral components consists entirely of narrow, ciliated, supporting cells, and forms the ciliated streak or Flimmerstreif. Coming between the Flimmerstreif and Drüsenstrief on each side is a tissue more nearly resembling the ordinary endoderm, and described by Professor von Heider (1895, p. 127 and fig. 16), as the "Entoderm-wucherung." I do not, however, regard it as homologous with the thickening of the mesenterial endoderm immediately behind the filament in its simple form as von Heider appears to (cf. his fig. 28).

Though differing in form the Zoanthean filament accords in histological detail with that of most other Actiniaria; the "Entodermwucherung," corresponding with what I have termed the "intermediate streak."

A peculiarity connected with the Zoanthean filament is that the Flimmerstreifen extend for some distance up each face of the perfect mesenteries, just
before the latter cease their connexion with the stomodæum ; the middle portion
is folded and in actual contact with the mesentery, while the two ends, or at any
rate the centrifugal end, may hang freely in transverse sections. The whole
structure has been denominated by Haddon and Shackleton (1891, p. 619), the
" reflected ectoderm," these authors regarding it as representing a portion of the
stomodæal ectoderm which has become transferred to the face of the mesenteries.
In the adult the reflected ectoderm and mesenterial filaments are always found in
absolute continuity with the ectoderm of the stomodæum. And it has also been
demonstrated by McMurrich (1891) and others, that even at an early stage in the
development of the embryo such a relationship can be recognized.

While not inclined to accept the ectodermal origin of the " reflected ectoderm," or of any portion of the mesenterial filaments, the former term may be
employed for the present as referring to parts now well known in Zoanthean
morphology. The Hertwigs (1879) first emphasized the fact that trifoliate
mesenterial filaments may appear on mesenteries of the lower orders which never
reach the stomodæum, and in all their structural details are indistinguishable
from those occurring on the first order ; and this appears to me to militate most
strongly against an ectodermal origin to any part of the Actinian filament. I
regard the continuity of the strongly ciliated stomodæal ectoderm, reflected
ectoderm, and the Flimmerstreifen and Drüsenstreif of the mesenterial filaments
as having a physiological rather than a morphological significance, as being
necessary, in fact, for the proper maintenance and regulation of the internal
circulation of the respiratory and digestive fluids in the mesenterial chambers of
and around the stomodæal region.

The histological characters of the tissues point to this, while the similarity of
structure is not so great as is sometimes assumed. The uniform nature of the
cells composing the Flimmerstreifen certainly contrasts strongly with the variety
met with in the stomodæal ectoderm, with the exception of those lining the
gonidial grooves. The grooves are usually more strongly ciliated, and but few
glandular or stinging cells occur amongst the supporting cells.

Less differences exist between the Drüsenstreif and stomodæal ectoderm,
while the Entodermwucherung shows no histological relationship with the latter.

The ciliation of the gonidial grooves, reflected ectoderm, and the Flimmerstreifen is more pronounced than that of any other region of the Actinian polyp,
and usually persists in preserved specimens even when not observable elsewhere.
The histological elements are also more specialized, pointing to a specialized
function. The cells are almost entirely of the extremely narrow ciliated type,
each with an oval-shaped nucleus, often larger in diameter than the cell itself,

and arranged at a different height in different cells in such a way that in sections they give rise to a very characteristic deeply-staining zone. Elsewhere in the polyp the association of histological elements is more varied, glandular cells or nematocyst-bearing cells mingling with supporting cells.

In two of the present species of *Parazoanthus*—*P. separatus* and *P. monostichus* —little or no reflected ectoderm is developed, and the mesenterial filaments are simple throughout, that is, only the middle lobe is present, not the lateral lobes. In longitudinal sections the ectoderm of the stomodæum is seen to be in continuity with the similarly deeply-staining tissue along the free edge of the mesenteries, but this is not continued for any distance up the faces of the latter; while transverse sections through the free edge of the mesenteries never present any structure which can be regarded as the Flimmerstreifen. *P. tunicans* exhibits on some of the mesenteries a weakly developed reflected ectoderm, and the filaments are trilobed for a very short distance below the termination of the stomodæum (Pl. xv., fig. 4).

In the figure which Haddon and Shackleton (1891, pl. lx., fig. 6) give of a transverse section through the terminal region of the stomodæum of *P. axinellæ*, the reflected ectoderm is strongly displayed, and on the free mesentery the filament exhibits the characteristic trifoliate appearance. In the genus Parazoanthus then every stage can be obtained in the presence or absence of the typical trifid Actinian filament, the variation evidently being dependent in some degree upon the dimensions obtained by the polyps.

The absence of the Flimmerstreifen from the mesenterial filaments is now known for several Actiniaria outside the Zoantheæ, and is the condition exhibited throughout the Madreporaria, as far as these have been studied. The character must be regarded as indicative of a lower degree of Actinozoan development, and in the two species of Parazoanthus mentioned, may be correlated with the very diminutive size of the polyps not necessitating the same vigorous internal circulation.

Professor Haddon and Miss Shackleton draw attention to the fact that the endoderm is often implicated in the upward reflection of the lower edge of the stomodæum. It is very noticeable in *Parazoanthus axinellæ*, the appearance in which species they figure. The same condition is also to be observed in all the species of Parazoanthus coming under my notice, as well as in many other Actiniaria and Madreporaria. In longitudinal sections it is evidenced by a strongly marked concave border to the mesentery as it leaves the stomodæum.

As the authors referred to remark, it has probably no morphological significance, and is no doubt exaggerated in a retracted state of the polyps.

The members of the genus exhibit a certain relationship in regard to the presence or absence of pigment granules and of zooxanthellæ. It is usually

2 E 2

noticed that where the former are present in excessive amount, the latter are absent, and *vice versâ* ; the two may, however, exist side by side in the same species. The granules are recognized as very small spheroidal bodies of various sizes, devoid of a nucleus and cell-wall, these being easily detected in the commensal algæ.

Most of the tissues of *P. Swiftii* are densely loaded with bright yellow granules of all sizes, but no zooxanthellæ occur.   The endoderm of *P. dichroicus* is also stated to be richly pigmented, and no zooxanthellæ are seen.   The converse holds in *P. tunicans*, the endoderm cells throughout contain an abundance of unicellular algæ, but pigment granules are practically absent ; *P. monostichus* and *P. separatus* show an admixture of granules and zooxanthellæ.   In the latter species a peculiar accumulation of brown pigment granules is found in the endoderm, about midway along the width of the mesenteries, this being the only occurrence in the polyp.

Similar relationships of granules and zooxanthellæ are afforded by other families of Actiniaria.   According to my observations pigment granules only are present in *Bunodes granulifera* and *B. Krebsii*, while they are replaced by zooxanthellæ in *Aulactinia stelloides*.   Most Sagartidæ contain zooxanthellæ, but in *Sagartia nivea* (Verrill), the substitution of granules has occurred.   The latter condition is also the case in *Actinoporus elegans* and in *Corynactis myrcia* already referred to.

It seems likely that in some cases the pigment granules may perform the same function as the commensal algæ—that of respiration.   If this be so, we may perhaps regard them as free chromo-plasts, aggregated in the other case within distinct cells, the zooxanthellæ.

Although the amount and relative proportion of the inclusions vary, yet a curious similarity in their nature holds throughout the genus.   Fine sand-grains and siliceous sponge spicules, with an occasional Radiolarian and Foraminiferal test, are characteristic of each species.   Carlgren found much the same in Gerardia.

Haddon and Shackleton note the inclusions to be fairly numerous in *P. anguicomus*, and less so in *P. axinellæ* and *P. dixoni*.   Calcareous sand-grains predominate in *P. tunicans*, and sponge spicules in *P. monostichus*, while both are numerous in *P. separatus*.   In the two latter the majority of the spicules are similar to those of the sponge with which the anemone is commensal.

A certain selection in the disposition of the foreign inclusions is also observable. Practically all the calcareous sand-grains of *P. tunicans* are limited to a narrow zone around the boundary of the ectoderm and mesoglœa ; while the sponge spicules are distributed throughout the middle layer, extending even to its inner boundary (Pls. XIII. and XIV., fig. 4).   Further, the spicules are most numerous in the capitular region, as is also the case with the two other species, rendering suitable sections in this region difficult to prepare. As the sponge spicules are very long in *P. mono-stichus*, much longer in fact than the thickness of the capitular wall, they are of necessity disposed in a regular circular series, very obvious in thick transverse sections (Pl. XIII., fig. 9).

The outer part of the column-wall of *P. Swiftii* is loaded with inclusions, but none extend beyond the encircling sinus. The mesoglœa there becomes extremely homogeneous in structure, an included cell even occurring but rarely. Such a strongly marked division of the mesoglœa into two parts—an outer, containing the inclusions, canals, cell-islets, etc., and an inner, practically homogeneous in nature and separated from the former by the encircling sinus—appears to be more or less general throughout the genus.

The size of the colonies and the extent to which the cœnenchyme is developed are likewise features of some importance. The simplest stage is exemplified by *P. separatus*, where each polyp is distinct and surrounded by only the merest trace of cœnenchyme. To include this exceptional instance, 1 have slightly added to the previous definitions of the genus. The few polyps in any colony of *P. monostichus* also afford but a bare indication of connecting cœnenchyme; while in a colony of *P. tunicans* or *P. dichroicus*, hundreds of polyps are, as it were, inserted in a common incrusting cœnenchyme. *P. Swiftii* is somewhat intermediate in the dimensions attained by its colonies and the amount of cœnenchyme produced. A few polyps only constitute a distinct colony, each arising from a clearly separable, though very limited, cœnenchyme.

The species I have examined, support the experience of Haddon and Shackleton (1891, p. 623), that "all the members of a single colony of diœcious Zoantheæ belong to the same sex." All the numerous polyps of *P. Swiftii* and *P. tunicans* sectionized were crowded with ova. It seems remarkable that of very many examples of *P. separatus* and *P. monostichus*, microscopically examined, none showed any trace of reproductive cells.

Parazoanthus tunicans, n. sp.

(Pl. x., fig. 11; Pl. XIII., fig. 7; Pl. xv., figs. 4, 5.)

Each colony consists of a thin cœnenchyme from which numerous polyps arise at short distances apart, the whole completely incrusting the main stems and smaller branches of a large Plumularia. On the smaller branches the polyps are arranged in a distichous manner, in a plane at right angles to that of the pinnulæ of the Hydroid, and the polyps on the two sides are either opposite or alternate. On the thicker stems their distribution becomes more irregular, and the polyps extend all round; they often arise obliquely to the surface of the cœnenchyme.

To the naked eye the surface of the cœnenchyme and column-wall is quite smooth, but with a lens minute white granulations—the foreign inclusions—are disclosed. The walls are thick and firm, but in some cases superficial wrinklings may be observed in preserved specimens.

The polyps were examined only in their retracted or partly retracted state. They are capable of complete retraction, in which condition they are usually mammiform; or they may be slightly longer, and flattened or rounded above, a small aperture remaining in the middle. Towards the base the column enlarges in diameter, especially in the most retracted individuals.

The capitular ridges are small, and can be distinguished and counted only with the assistance of a lens; they are wedge-shaped and acute, and vary in number from 14 to 16. The ridges and furrows are most distinctly indicated during partial extension.

The tentacles are short, apparently rounded at their apex, and dicyclic, fourteen to sixteen occurring in each cycle. The mouth is rounded or slit-like, and the lips prominent.

The colour of the cœnosarc and column-wall is greyish, being determined by that of the included particles ; the tips of the capitular ridges are a little lighter ; the tentacles and disc are brown.

The height of retracted polyps above the cœnenchyme is about 2 mm., and the diameter the same.

ANATOMY AND HISTOLOGY.

The ectoderm of the column-wall is a continuous layer, that is, it is not broken up by crossing strands of mesoglœa, as is the case in many Zoantheæ. Superficially, it is devoid of any recognizable cuticle or sub-cuticle, and the constituent cells are more rounded than columnar in outline. The internal limitations of the layer are very irregular and indeterminate in places, most of the inclusions occurring around its boundary with the mesoglœa, while cells pass from it into the mesoglœa (Pl. XIII., fig. 7 ; Pl. XV., fig. 4).

Small colourless nematocysts occur, but are not very numerous.

The mesoglœa is moderately thick, and near its internal border contains a narrow, interrupted, encircling sinus filled with cells closely resembling those of the ectoderm ; in the distal region, where the sinus becomes broader, it includes numerous nematocysts.

Isolated cells and cell-islets are scattered throughout the mesoglœa; and a few siliceous sponge spicules are included, in addition to the predominating calcareous sand-grains. The latter are very small and practically limited to its peripheral border ; they are dissolved out by acids leaving only irregularly-shaped lacunæ. In regard to the foreign inclusions, a decided selection is manifested in that the calcareous incrustations are limited towards the periphery of the wall, while the siliceous sponge spicules and a few Radiolarian tests are more internal and more distal.

The endoderm cells are loaded with zooxanthellæ. Proximally there is only a faint indication of an endodermal circular musculature, but towards the apical region the fibres become stronger and more concentrated, and constitute a weak, diffuse, endodermal sphincter muscle, the mesoglœa forming deep, closely-arranged bays for its reception. No part of the muscle, however, becomes actually enclosed in mesoglœa, except at the places where the mesoglœa of the mesenteries is united with that of the column-wall (Pl. XIII., fig. 7).

The cœnosarc surrounding the Hydroid stem, and connecting one polyp with another, contains inclusions similar to those of the column-wall. Irregular channels with a thick lining of endoderm serve as a means of communication between the cœlenteron of one polyp and another. In sections the Hydroid stem is completely embedded in mesoglœa; this latter also contains abundant cells and cell-islets.

The ectoderm of the tentacles discloses a peripheral zone of small, narrow nematocysts throughout its length. The mesoglœa is thin and very slightly plaited for the support of a weak ectodermal and endodermal musculature, and a nervous layer connected with the former is distinguishable. The endoderm is loaded with zooxanthellæ, and completely fills the lumen in contracted tentacles.

The disc is extremely thin-walled, but becomes a little thicker near the tentacular region, where nematocysts and gland cells occur.

The stomodæum is of small vertical extent. A single gonidial groove is indicated, very shallow in some examples and deeper in others, while the walls rarely display any vertical folding. The ectoderm is constituted of the usual ciliated supporting cells, granular gland cells, and but few nematocysts; in the region of the groove, glandular cells are very scarce. An ectodermal and endodermal musculature can be made out, though but feebly developed; an ectodermal nervous layer is also displayed. The mesoglœa is much thinner than the ectoderm, and undergoes no additional thickening at the groove. The endoderm is broad, and its cells contain many zooxanthellæ.

At its lower termination the wall of the stomodæum is backwardly and outwardly directed for a short distance; and the ectoderm is in continuity with the tissue of almost exactly similar nature which runs radially along the edge of the perfect mesenteries, and, as the "Reflected ectoderm," passes for a very short distance up each face of the perfect mesenteries.

The reflected ectoderm is not developed to the same extent on all the mesenteries, and very rarely presents a similar appearance on the two faces of the same mesentery. It is constituted of extremely narrow ciliated supporting cells, the oval nuclei of which are arranged at different heights in the different cells, so that a distinct, deeply-staining, nuclear zone is produced in sections (Pl. xv., fig. 4).

In transverse sections, a little above the stomodæal termination, the reflected ectoderm is thrown into a few short vertical folds, and centrifugally is free from the mesentery, while centripetally it can be traced in continuity with the ectoderm of the stomodæum.

This continuity is, however, not one of exactly similar tissues throughout, but is interrupted at places by a tissue of a different nature. The cell nuclei of this are rounded, and not arranged in a distinct zone, and the whole stains less deeply and contains zooxanthellæ and granular gland cells. To this tissue, as met with in *Zoanthus chierchiæ*, von Heider (1895, p. 129) has applied the term Drüsenwulst." Though its presence can be easily recognized in *P. tunicans*, it is not so well developed as in the genus *Zoanthus*, where, owing to the increased size and length of the polyps, the reflected ectoderm and mesenterial filaments are better displayed and more favourable for study.

As shown in Pl. xv., fig. 4, a filament is a complex structure, the sagittate or lanceolate form in transverse sections being characteristic of the Zoantheæ. The outer border of the lateral lobes is constituted of ciliated, extremely narrow cells, the associated nuclei forming a very regular, densely-staining zone. It is to this tissue that I consider the term ciliated streak or Flimmerstreif should be restricted, and not applied to the lateral lobe as a whole, as is usually done. The inner layer of the lobes is formed of endoderm cells, indistinguishable from the epithelium of the mesentery. The tissue occurring around the termination of the middle lobe is made up of ciliated supporting cells, granular gland cells, and nematocysts. The term glandular streak, Drüsenstreif, or Nesseldrüsenstreif should, in my opinion, be employed only for this part of the middle lobe, the intermediate streak coming between it and the ciliated streak. Towards its termination the mesoglœa is slightly swollen, and delicate muscular fibres border it anteriorly.

The mesenteries exhibit the macrocnemic arrangement, that is, the dorsal, or sulcular pair of imperfect directives has a pair of mesenteries on each side— of which the first is a perfect mesentery, and the other an imperfect—and the succeeding pair consists of two perfect mesenteries (Pl. xv., fig. 4). Beyond these the pairs consist of an imperfect and a perfect mesentery until the neighbourhood of the ventral or sulcar directives is reached, when the arrangement in pairs becomes a little irregular, this being the region in which new mesenteries are added. In one polyp eight perfect mesenteries occurred on each side, while in another eight were present on the left and six on the right side.

The mesenterial musculature is extremely feeble, and the parieto-basilar is clearly distinguishable. The mesoglœa is broad at its origin in the column-wall, but narrows rapidly beyond. A very short basal canal and several cell-islets occur in the expanded portion. The mesenterial endoderm is broad and loaded with zooxanthellæ; nematocysts also occur sparingly.

Male gonads were present in all the numerous polyps sectionized from the one colony. The surrounding mesenterial epithelium is enormously thickened and the ripe spermaria are enclosed in the very thin mesogloea. Around their margin are the deeply-staining sperm mother-cells; filling the greater part of the interior are the heads of the ripe spermatozoa, while towards one side are aggregated the tails of the spermatozoa (Pl. xv., fig. 5).

One very large, much-branched colony was trawled from a depth of 10–14 fathoms on the Pedro Bank, 11th April, 1898, incrusting an aborescent *Plumularia*, as much as 100 cm. high. The coenenchyme was continuous nearly throughout the surface of the Hydroid, only the smallest terminations being free.

Apparently no Zoanthid at all resembling this form has been described from Antillean waters, nor as a member of the nearly related Actiniarian fauna on the western coast of Central and South America.

In habit and external features it compares most closely with *Parazoanthus dichroicus*, Hadd. and Shack. (1891, p. 698), obtained by Prof. Haddon from Torres Straits, incrusting a specimen of *Plumularia ramsayi*. It thus forms another instance of the strong relationship, particularly in the Zoantheæ, which is being established between the Australian and Caribbean seas.

The capitular ridges in *P. dichroicus* are about eighteen, an increase of two or three beyond the number prevailing in *P. tunicans*.

I have never observed any dichroic effect given to the alcohol from preserved material, a peculiarity emphasized in the specific name of the former. Histologically important differences are indicated, which leave no doubt as to the distinctness of the two species.

The incrusting particles of sand are siliceous in the older species and calcareous in the new; the encircling sinus is filled with dark-brown granular pigment in the one, but not in the other. The latter distinction is associated with the absence of zooxanthellæ in the pigmented form, while they are abundant in *P. tunicans*, which, conversely, is devoid of pigment. The mesenterial musculature is less developed in the last-mentioned species.

External characters alone readily separate it from all other known species of Parazoanthus.

Parazoanthus separatus, n. sp.

(Pl. x., figs. 12, 13; Pl. xiii., fig. 8; Pl. xiv., fig. 4.)

In their retracted condition the isolated polyps present themselves as small, circular, light-coloured discs, or low mammiform prominences, distributed with considerable regularity over the surface of massive, dark-coloured sponges. What may be regarded as a very narrow border of cœnenchyme surrounds each individual polyp. Only occasionally are two or more polyps still united as a result of reproduction by budding, and all stages in the separation of one from the other can be observed. A thin cœnenchyme connects two polyps before their isolation, but chain-like colonies are never produced, as in the next species. The column-wall is smooth, but with a lens, minute, white, incrusting granulations are seen. These give a certain rigidity to the polyps, so much so that, in preserved material, the walls readily split in two.

Retraction is complete in most examples ; only a very small circular depression remains above, not sufficiently large to allow the mouth or disc to be seen. The capitular ridges are wedge-shaped, and number from twelve to sixteen, twelve being the most usual. The tentacles are extremely short, and, as seen in sections, are almost invariably twenty-four, arranged in two cycles. The disc is circular and semi-transparent, and exhibits radiating grooves corresponding with the internal attachment of the mesenteries. The peristome is raised, the mouth slit-like.

The cœnenchyme and column are dull white, due to the numerous included calcareous sand-grains; the tentacles and disc are dark brown.

The diameter of retracted polyps is about 2·5 mm., the height 1·5 mm.

Anatomy and Histology.

All that portion of the wall of the polyp which is embedded in the sponge may be regarded as the base, and discloses a different histological character from that of the free column-wall. It is convex in outline, but somewhat flattened and expanding peripherally, and is sharply marked off from the surrounding sponge tissue, showing that there is no intimate cellular relationship between the two. A very thin cuticle can also be traced (Pl. xiv., fig. 4).

The ectoderm in places is not readily distinguishable from the mesoglœa, the latter layer being so crowded with cells as to render the ground substance almost unrecognizable (Pl. xiii., fig. 8). The ectodermal cells are large and more or less spherical in outline, not forming a columnar epithelium; their protoplasm stains very strongly.

The mesoglœa is broad, and, as a whole, stains very deeply, the result of the presence of the cellular constituents, mostly in the form of cell-islets. Cells are included to an extent greater than I have met with in any other Actinian; they are all large, and the cytoplasm, in addition to the nucleus, readily takes up any stain. The matrix is scarcely discernible except at the inner border, where it may sometimes be observed in connexion with a mesentery. Sponge spicules are present in both the ectoderm and mesogloea, though not to the same extent as might have been expected from the nature of the commensalism. Lacunæ also occur in decalcified specimens, indicating where calcareous sand-grains had been included. The endoderm is a broad layer, crowded with zooxanthellæ. No basal musculature has been detected.

At the boundary of the base and column occurs an expansion of the wall, there being, as noticed amongst the external characters, a slight formation of coenenchyme (Pl. xiv., fig. 4).

The ectoderm of the column-wall is a broad continuous layer, the columnar character of the cells not being clearly indicated in sections. Its internal limits are ill-defined, partly owing to the foreign inclusions tending to break up the layer, and also to the fact that abundant cells pass from it into the mesogloea. A small oval nematocyst, which does not stain, is scattered sparingly and irregularly throughout.

The mesogloea contains small, rounded or elongated cells with granular proto-plasm, and also cell-islets, not, however, to the same extent as in the mesogloea of the base. Towards its internal border a very irregular, narrow, encircling sinus occurs, and beyond this it is much more homogeneous. Owing to the strongly cellular nature of the outer region of the mesogloea the encircling sinus is not so distinct as in most of the species investigated by Haddon and Shackleton, nor as in *P. tunicans*, where the mesogloeal matrix is much more uniform. The cells included in the sinus possess very granular protoplasm, and abundant nema-tocysts similar to those in the ectoderm; these latter are particularly numerous in the distal region of the polyps.

Numerous cellular connexions can be traced between the irregular internal limits of the ectoderm and those of the encircling sinus.

The whole mesogloeal layer contains foreign inclusions, more abundant, how-ever, peripherally; they are mainly calcareous sand-grains which are dissolved out by acids. Silicious sponge spicules are particularly abundant in the upper region of the column, and always remain in microscopic preparations. The sand-grains are more restricted in their distribution to the region of contact of the ectoderm and mesogloea.

The endoderm of the column-wall resembles that of the base, but above is much thickened between the mesenteries, while it is narrow below. An extremely weak, endodermal musculature extends the whole length of the column. At the capitulum, the mesogloea becomes sinuous in sections, and the muscle fibres are here a little stronger and represent the sphincter muscle (Pl. xiv., fig. 4).

The sphincter is of a diffuse endodermal character; the mesogloeal folds

2 F 2

are rather narrow and deep, but the presence of numerous sponge spicules interferes with a detailed study. As noticed amongst the external characters the polyps are capable of excessive retraction, so much so as to obliterate a great part of the cœlenteric space, and produce a great displacement of the disc and stomodœal walls.

The ectoderm of the tentacles presents throughout its extent a peripheral layer of small, narrow nematocysts, differing from the oval form in the column-wall. Below this nematocyst layer a nuclear zone is usually separable from the more internal nervous and muscular elements. An ectodermal and an endo-dermal musculature are developed; the former much the stronger. The layer of nerve fibrils is often distinguishable in connexion with the ectoderm. The mesoglœa is very thin, but a little better developed proximally; it is finely plaited for the support of the ectodermal musculature.

The endoderm contains zooxanthellæ, and very often fine pigment granules; these latter are also found in the ectoderm. Spicules occur in some abundance in the tentacular tissues, somewhat more numerous in the outer than in the inner cycle. Though such a position for inclusions is exceptional they are met with in all the examples studied, and in such a manner as to leave little doubt that they are not the result of displacement during the preparation of the sections.

The disc is very thin, and peripherally closely resembles the tentacles in structure; a few nematocysts occur in the ectoderm, as well as numerous deeply-staining granular gland cells.

The vertical height of the stomodæum is remarkably small in contracted specimens (Pl. xiv., fig. 4) ; and in a series of transverse sections the sulcar end terminates in advance of the sulcular. The single gonidial groove is clearly indicated. In transverse sections the wall is usually cut through twice as a consequence of the partial reflection of the internal termination of the stomo-dæum. As seen in the figure, the stomodæal wall passes slightly upwards and outwards for a considerable distance. The ectoderm displays the usual histo-logical structure, consisting mainly of ciliated supporting cells, the combined nuclei of which give rise to a very distinct zone ; granular gland cells, which also stain deeply, are abundant, especially in the upper regions, but nematocysts do not appear to be developed. No ectodermal musculature is discernible over any part of the stomodæum. The mesoglœa is very thin, and undergoes no appreci-able thickening at the groove. In vertical sections the stomodæal ectoderm is in continuity with the mesenterial filaments; but there is no special forma-tion of reflected ectoderm.

Owing to the extreme retraction and the shortness of the stomodæum, some difficulty is experienced in making out the arrangement of the mesenteries; but out of numerous examples sectionized, I have been able to definitely ascertain the macrocnemic arrangement in several.

The mesenteries below the stomodæum are very short in transverse sections, and extend but a little distance vertically; two or three are continued for some way below the others, but which are in relation to the directives could not be determined. As the free edge of the mesentery leaves the stomodæum it becomes deeply concave. Owing to this, and the shortness of the stomodæum, the perfect mesenteries in transverse sections of some retracted polyps appear free even before the stomodæum is reached, one half being still connected with the concave disc, and the other with the column-wall, each with the filamental tissue at its free termination.

The endodermal epithelium of the mesenteries resembles that of the body-wall, and contains many zooxanthellæ.

In the upper region of the polyps, the mesenterial mesogloea as it leaves the column-wall is much and irregularly thickened, and contains cell-islets, but beyond this the layer is extremely thin. There is a distinct indication of a parieto-basilar muscle on each side, but the longitudinal musculature is not sufficiently developed to allow of a study of the paired arrangement of the mesenteries being made.

No reflected ectoderm occurs on any of the mesenteries. In transverse sections around the termination of the stomodæum an appearance of such is presented, but it is merely the Drusenstreif which here runs horizontally. The tissue is never folded, as is usually the case, with the reflected ectoderm, while granular cells and nematocysts are present in addition to the supporting cells. Vertical sections also reveal a similar absence.

For some little distance from their origin at the stomodæum, the filaments in section have an irregular outline. They are simple throughout their length, consisting only of the middle lobe or Drüsenstreif. In the lowermost region the mesenteries may divide at their free edge into three branches, each capped by a filament which is cordate in transverse section. The latter is sharply cut off from the rest of the mesentery, and stains much more deeply. The mesenterial endoderm is usually thickened immediately behind the filament, in some cases partly surrounding the filament; otherwise it differs in no important respect from the remaining mesenterial epithelium.

No gonads were present in any of the numerous examples sectionized.

The form described above was trawled on several occasions from a depth of 10–14 fathoms on the Pedro Banks, Caribbean Sea, embedded in the superficial tissues of some massive, dark-coloured sponges. From the number of sponges trawled, each bearing the commensal Zoanthid, the species must be very abundant in these regions. The individual polyps are closely scattered over the whole superficial area of the sponge, and are arranged with considerable regularity as regards distance apart. They are generally a little closer and less regular in the lower region of the sponge, where growth cannot be proceeding so rapidly as more distally. Numerous small pores are distributed over the sponge, and in most of these a commensal Alphæus was found.

I have hesitated for some time as to whether the form can be referred to any of the known Antillean sponge-incrusting species. The one most likely is *Zoanthus parasiticus*, D. & M., in which the polyps are isolated.

This is, however, stated to be a veritable Zoanthus with fleshy walls, not hardened by fleshy inclusions. Under these circumstances I think it is best to regard the species as distinct, awaiting further discoveries to indicate the true nature of the Zoanthus. The Caribbean Sea is obviously very rich in examples of anemones commensal with sponges, but this and the next described species, along with *P. Swiftii*, are readily distinguished both externally and anatomically.

Parazoanthus monostichus, n. sp.

(Pl. x., fig. 14; Pl. xiii., fig. 9.)

The polyps give rise to extremely small colonies embedded in the superficial tissues of a dark purplish sponge, over the whole of which they are distributed with considerable regularity. In the retracted condition of the polyps, the colonies appear as minute, light-coloured catenulations, contrasting strongly against the dark sponge. From two to seven or eight individual polyps are associated in a single row, but sometimes one or two may be produced laterally, and so give rise to an irregularly-shaped colony. Rarely the polyps are isolated. An extremely narrow border of cœnenchyme surrounds each colony or individual. Multiplication takes place by budding, and the individuals are often so closely contiguous that no intervening cœnenchyme is apparent. All stages in the separation of one polyp from another can be observed, the cœnenchyme becoming drawn out more and more until the constriction breaks down. In retraction the polyps are flattened and scarcely raised above the general surface of the sponge. They appear to be incapable of complete retraction; the capitulum is always fully visible, and a wide apical aperture remains in most, so that the mouth and central part of the disc are exposed. The capitular ridges are wedge-shaped, and number about 10. The surface is smooth, but minute, opaque white particles are embedded in the capitular region.

The polyps have been observed only in the semi-retracted condition, so that no details of the external appearance of the tentacles and disc can be added. The mouth is slightly oval.

In preserved specimens the cœnenchyme and column are a dull white, due to the included particles.

The diameter of the retracted polyp is about 1 mm., and the height 0·5 mm. The species is probably the smallest Actinian known.

ANATOMY AND HISTOLOGY.

Owing to the exceptional smallness of the polyps and the inclusion of numerous large, silicious, sponge spicules, the anatomical study of the species is carried out under considerable difficulties, and characters of fundamental import-ance, such as the arrangement of the mesenteries and the nature of the sphincter muscle, remain in some uncertainty.

The ectoderm of the base is in contact with the tissues of the sponge on the one hand, and on the other is scarcely distinguishable from the outer part of the mesoglœa, numbers of its cells passing into the latter. The individual cells are not disposed to form a columnar epithelium, as is usually the case, but are rounded or irregular in shape, and both the nucleus and the cytoplasm stain deeply.

The mesoglœa is divisible into two portions : an outer, broader part, much broken up by sinuses and cell-islets ; and an inner, narrow, limiting part, more uniform in structure, and thickening along the line of attachment of the mesenteries. The former broadens much in some regions, and the large individual cells of the cell-islets, all with deeply-staining contents, become more distinct from one another. The endoderm is a somewhat thick layer, and contains abundant zooxanthellæ.

The outline of the base is convex, and in vertical sections across the length of the colonies the proximal region of the wall of the polyp is a little expanded laterally, constituting a narrow cœnenchyme. The mesoglœal layer here becomes thickened, and many silicious sponge spicules are included.

Large, perfect sponge spicules, arranged very closely in a circular manner, are particularly numerous in the capitular region of the column-wall (Pl. XIII., fig. 9), while calcareous sand-grains are scarce.

The ectoderm of the column-wall is a layer of non-columnar cells, and medium-sized, colourless, oval nematocysts are abundant, especially in the more distal regions. A cuticle is also observable. The internal limitations of the ecto-derm are irregular, the layer passing more or less insensibly into the mesoglœa. The latter is so crowded with cells, that it stains nearly as deeply as the ectoderm. Large cell-containing spaces, connected with the ectoderm, probably represent the encircling sinus characteristic of the genus, and met with in the two previous species, but, owing to the numerous inclusions, it is impossible to make out the relations of one cavity to another. More internal than these cell-spaces the mesoglœa is practically homogeneous, and affords a sharp boundary line with the endoderm.

The endoderm is broad and contains numerous zooxanthellæ ; its cells are much elongated in the narrow mesenterial spaces. Only faint indications of an endodermal circular musculature can be made out. It is unfortunate that in the capitular region, where the sphincter muscle should occur, the walls are so thin and the large sponge spicules so closely aggregated, as to render suitable sections a matter of practical impossibility. The delicate walls in every case readily break away with the inclusions.

From a knowledge of related forms it can, I think, be safely inferred that if a sphincter is developed it will be of an endodermal, diffuse, and extremely weak type. The power of retraction is not possessed to the same degree as in the previous species.

The ectoderm of the tentacles is a broad, columnar layer ; small, narrow nematocysts occur peripherally, and occasionally one of the larger oval forms similar to those in the column-wall. The merest traces of an ectodermal and also of an endodermal musculature can be detected. The mesoglœa is extremely narrow ; the endoderm is loaded with zooxanthellæ and with deeply-staining granules of various sizes, and, in retracted examples, completely fills the lumen.

The disc in all its three layers is a very thin structure. In retracted specimens it is deeply concave outwardly, affording space above for the tentacles, while below it comes almost in contact with the floor of the cœlenteron, nearly obliterating the cœlenteric cavity. In consequence of the extreme shortness of the polyps as a whole, and of this approximation of the disc and base, the study of the paired arrangement of the mesenteries is almost fruitless.

The peristome remains elevated, and the stomodæum is comparatively large in sections. As shown in Pl. XIII., fig. 9, transverse sections pass at the same time through the column-wall, tentacles, elevated peristome, and stomodæum ; only exceptionally can a mesentery be traced from the column-wall to the stomodæum.

The stomodæum is usually oval-shaped in tranverse sections, and the single gonidial groove is clearly indicated ; the lateral walls may be thrown into a few vertical folds. The ectodermal epithelium consists mainly of ciliated supporting cells with nematocysts and gland cells ; the two latter are practically absent at the sulcar end where the groove occurs. Large granular gland cells, the contents of which do not stain in borax carmine, are also met with. The mesoglœa is a very thin layer, thickening somewhat at the groove ; the endoderm is broad and crowded with zooxanthellæ.

In vertical sections the stomodæal ectoderm is seen to be in continuity with the filamental tissue of the mesenteries. The lower termination of the stomodæum is folded backwardly and outwardly, so that in transverse sections through this region its endoderm and ectoderm are cut through twice ; and, further, the ectoderm appears to be continued radially for some distance along the mesenteries. The mesenterial filaments are, however, simple throughout, that is, only the middle lobe is present, the lateral lobes with the Flimmerstreifen not being developed.

Although numerous polyps have been sectionized, it has been found impossible to make out the complete arrangement of the mesenteries. In most ten mesenteries are perfect, being united with the stomodæum at varying intervals. Nine perfect mesenteries occurred in one example. No certain indications of imperfect mesenteries were afforded. The mesoglœa of the mesenteries is swollen at its origin in the column-wall and encloses cell-islets ; beyond the origin it thins rapidly. An extremely weak parieto-basilar musculature occurs. The retractor muscle fibres of the mesenteries are similarly very feeble, the mesoglœa being slightly plaited to give increased support; the fibres appear to be strongest about the middle of the width of the mesentery. The mesenterial endoderm is broad and crowded with zooxanthellæ; it is more swollen below, and contains an abundance of small, spherical, apparently non-nucleated bodies, which stain deeply. Pigment granules are practically absent, but about the middle of transverse sections of mesenteries a peculiar accumulation of fine, yellowish brown granules occurs on each face, very limited in its radial extent. The endoderm throughout the polyps occasionally contains large zooxanthellæ-like bodies with a highly refractive cell-wall. No gonads were indicated in any of the polyps sectionized.

The species was trawled on only one occasion, February 10, 1898, at a depth of 10–14 fathoms, on the Pedro Bank, Caribbean Sea, commensal with a silicious sponge.

In a list of the Actiniaria around Jamaica (1898*a*), I identified the form as the much debated *Bergia catenularis*, Duch. & Mich., its commensal habit and very decided catenulariform appearance suggesting this species most forcibly. I am now convinced, however, that the safest course is, for the present, to regard it as a new species and await the possibility of discovering others which may approach the older species more closely. Especially may this be the case in regard to the nature of the cœnenchyme connecting the individual polyp. From Duchassaing and Michelotti's figure, *B. catenularis* appears to have this better developed than in the present species, and more in the form of stolons, but I do not attach much importance to the statement that the connexions arise from the upper part of the polyps. This is probably merely a result of the colonies being partly embedded in the sponge, and *P. monostichus* affords indications of the same feature. There is little doubt that the two species of Bergia will, when rediscovered and sectionized, be found to belong to the genus Parazoanthus.

It is evident that we have in West Indian waters many species of small Zoanthids commensal with sponges; and, where such very few external characters are available for diagnostic purposes, their identification with previously described forms must be a matter of some uncertainty, unless one has the actual types or a number of the different species for comparison. When this can be done it will be found much easier to unite supposed new forms with the older than to rectify the confusion of synonymy.

Now that a few more representatives of these commensal anemones are available for comparison the present species appears to agree more closely with Duchassaing and Michelotti's *Gemmaria Swiftii* than does the form I identified as such in the former paper on the Jamaican Zoanthcæ.

REFERENCES.

1767. ELLIS, J.:
    "An account of the Actinia sociata, etc."—Phil. Trans., vol. lvii.

1786. ELLIS, J., and SOLANDER, D.:
    "The Natural History of many curious and uncommon Zoophytes collected from various parts of the Globe."—London.

1788. GMELIN, J. F.:
    In Linnæus' "Systema naturæ."—Edit. xiii., Lipsiæ.

1817. Lesueur, C. A.:
    "Observations on several species of the genus Actinia. Illustrated by figures."—Jour. Acad. Nat. Sci., Philadelphia, vol. i., pp. 169–189.

1834. Ehrenberg, C. G.:
    "Die Korallthiere des Rothen meeres."—Berlin.

1846. DANA, J. D.:
    "Report on Zoophytes."—U. S. Explor. Exped., 1838–1842.—Philadelphia.

1846. ALLMAN, G.:
    "Description of a new genus of Helianthoid Zoophytes (Corynactis)."—Ann. Mag. Nat. Hist., vol. xvii.

1847. JOHNSTON, G.:
    "A History of British Zoophytes," 2nd Edit.—London.

1850. DUCHASSAING, P.:
    "Animaux Radiaires des Antilles."—Paris.

1857. MILNE-EDWARDS, H.:
    "Histoire Naturelle des Coralliaires or Polypes proprement dits," vol. i.—Paris.

1860. GOSSE, P. H.:
    "Actinologia Britannica. A History of the British Sea-Anemones and Corals."—London.

1860. DUCHASSAING, P., ET MICHELOTTI, J. :
  " Mémoire sur les Coralliaires des Antilles."—Mem. Reale Accad. Sci. Turin, Ser. II.,
    Tom. XIX.

1866. DUCHASSAING, P., ET MICHELOTTI, J. :
  " Supplément au Mémoire sur les Coralliaires des Antilles."—Mem. Reale Accad. Sci. Turin,
    Ser. II., Tom. xxiii.

1869. VERRILL, A. E. :
  " Notes on Radiata. Review of the Corals and Polyps of the West Coast of America."—
    Trans. Connect. Acad., vol. i., 1868-1870.

1877. KLUNZINGER, C. B. :
  " Die Korallthiere des Rothen Meers. Theil I., Die Alcyonarien und Malacodermen."—Berlin.

1877. MOSELEY, H. N. :
  " On new forms of Actiniaria dredged in the deep sea ; with a description of certain Pelagic
    surface swimming species."—Trans. Linn. Soc., 2 ser. Zool., vol. i.

1879. HERTWIG, O. u. R. :
  " Die Actinien anatomisch und histologisch mit besonderer Berückshictigung des Nerven-
    muskelsystems untersucht."—Jena.

1882. HERTWIG, R. :
  " Report on the Actiniaria dredged by H. M. S. ' Challenger ' during the years 1873-1876."
    —Zoology, vol. vi.

1883. ANDRES, A. :
  " Le Attinie."—Atti. R. Accad. dei Lincei, ser 3a, vol. xiv. (also in " Fauna u. Flora d.
    Golfos v. Neapel," in , Leipzig, 1884).

1888. HERTWIG, R. :
  " Report on the Actiniaria dredged by H. M. S. ' Challenger ' during the years 1873-1876."—
    Supplement. Zoology, vol. xxvi.

1889. M‘MURRICH, J. P. :
  " The Actiniaria of the Bahama Islands, W. I."—Journ. Morphol., vol. iii., No. 1.

1889 a. M‘MURRICH, J. P. :
  " A contribution to the Actinology of the Bermudas."—Proc. Acad. Nat. Sci., Philadelphia.

1890. MITCHELL, P. C. :
  " *Thelaceros rhizophoræ*, n. g, et n. sp., an Actinian from Celebes." Quar. Jour. Micr. Sci.,
    (N.S.), vol. xxx.

1891. M‘MURRICH, J. P. :
  " Contributions on the Morphology of the Actinozoa. II. On the development of the
    Hexactiniæ."—Jour. Morph., vol. iv., No. 3.

1891. CARLGREN, O. :
  " *Protanthea simplex*, n. gen. u. sp., eine eigentümliche Actinie, Vorl. Mitteilung." Öfversigt
    Kongl. vet.-Akademiens Förl., No. 2.

1891. HADDON, A. C., and SHACKLETON, ALICE M. :
  " A Revision of the British Actiniæ. Part II. : Zoantheæ."—Trans. Roy. Dublin Soc., vol. iv.
    (ser. II.).

208   J. E. DUERDEN—*Jamaican Actiniaria:*

1891a. HADDON, A. C., and SHACKLETON, ALICE M. .
    " Reports on the Zoological Collections made in Torres Straits by Prof. A. C. Haddon, 1888-1889. Actiniæ: I. Zoanthem."—Trans. Roy. Dublin Soc., vol. iv. (ser. II.).

1892. SIMON, J. A. :
    " Ein Beitrag zur Anatomie und Systematik der Hexactinien. Inaug. Dissertation."—München.

1893. M'MURRICH, J. P. :
    " Report on the Actiniæ collected by the U. S. Fish. Commission Steamer ' Albatross ' during the winter of 1887-1888."—Proc. U. S. National Mus., vol. xvi.

1893. CARLGREN, O. :
    " Studien über Nordische Actinien, I."—Kongl. Svenska vet.-Akademiens Handl., Bd. 25.

1893. SAVILLE-KENT, W. :
    " The Great Barrier Reef of Australia."—London.

1895. CARLGREN, O. :
    " Ueber die Gattung *Gerardia*, Lac.-Duth."—Ofversigt af Kongl. Vet.-Akademiens Förhandlingar, No. 5.

1895. VON HEIDER, A. R. :
    " Zoanthus chierchiæ, n. sp."—Arbeit. Zool. Inst. Graz., v.

1896. M'MURRICH, J. P. :
    " Notes on some Actinians from the Bahama Islands, collected by the late Dr. J. I. Northorp." Ann. N. Y. Acad. Sci., vol. ix.

1896. HADDON, A. C., AND DUERDEN, J. E. :
    " On some Actiniaria from Australia and other Districts."—Trans. Roy. Dublin Soc., vol. vi. (ser. II.).

1896. KWIETNIEWSKI, C. R. :
    " Revision der Actinien, welche von Herrn Prof. Studer auf der Reise der Korvette Gazelle um die Erde gesammelt wurden."—Jenaische Zeitschr., Bd. xxx.

1896. KWIETNIEWSKI, C. R. :
    " Actiniaria von Ternate." Abhandl. d. Senckenb. Naturf. Ges., Bd. xxiii.

1897. DUERDEN, J. E. :
    " The Actiniarian Family Aliciidæ." Ann. Mag. Nat. Hist., Ser. 6, vol. xx.

1898. DUERDEN, J. E. :
    " Jamaican Actiniaria. Part I. : Zoanthem. Trans. Roy. Dublin Soc., vol. vi. (ser. II.).

1898. DUERDEN, J. E. :
    " The Actiniaria around Jamaica."—Journ. Instit. Jamaica, vol. ii., No. 5.

1898. KWIETNIEWSKI, C. R. :
    " Actiniaria von Ambon und Thursday Island."—Jenaische Denkschriften, viii.

1898. HADDON, A. C. :
    " The Actiniaria of Torres Straits."—Trans. Roy. Dublin Soc., vol. vi. (ser. II.).

1898. DUERDEN, J. E. :
    " On the Relations of certain Stichodactylinæ to the Madreporaria."—Jour. Linn. Soc., vol. xxvi.

EXPLANATION OF PLATE X.

PLATE X.

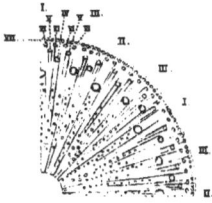

1.

2.

3.

4.

5.

6.

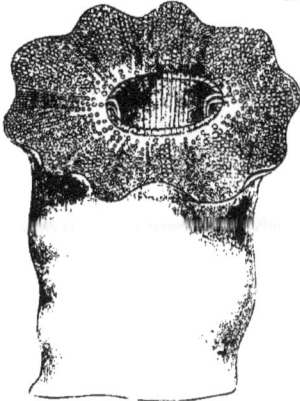

7.

8.

9.

10.

11.

12.   13.

14.

J. E. D. del.

EXPLANATION OF PLATE XI.

PLATE XI.

col. w.,	column wall.	*m.,*		mesentery.
c. is.,	cell-islet.	*m. fil.,*		mesenterial filament
c. s.,	ciliated streak (Flimmerstreif).	*mes.,*		mesoglœa.
cu.,	cuticle.	*mg. st.,*		marginal stoma.
d.,	directives.	*mr. m.,*		macrocnemic mesentery.
disc,	disc.	*nem.,*		nematocyst.
ect.,	ectoderm.	*nr. l.,*		nerve layer.
ect. can.,	ectodermal canal.	*ob. m.,*		oblique muscle.
ect. m.,	ectodermal muscle layer.	*p. b. m.,*		parieto-basilar muscle.
enc. s.,	encircling sinus.	*per. st.,*		perioral stoma.
end.,	endoderm.	*r. ect.,*		reflected ectoderm.
end. m.,	endodermal muscle layer.	*r. m.,*		retractor muscle.
fb. l.,	fibrillar layer.	*r. s.,*		reticular streak.
fos.,	fossa.	*s. d.,*		sulcar directives.
gl. c.,	gland cell.	*sl. d.,*		sulcular directives.
g. gr.,	gonidial groove.	*s.p.,*		spermarium.
g. s.,	glandular streak.	*sph. m.,*		sphincter muscle.
	(Drüsenstreif, Nesseldrüsenstreif).	*st.,*		stomodæum.
inc.,	incrustations.	*tent.,*		tentacle.
i. s.,	intermediate streak.	*zoox.,*		zooxanthellæ.
lac.,	lacunæ.	*I., II., III., &c.,*		orders of mesenteries or tentacles.

Figure

1. *Phymanthus crucifer.* Transverse section through a portion of the column-wall and a mesentery of the first order, and one of the mesenteries of the fourth order, taken a little below the stomodæal region. × 320.

2. *Phymanthus crucifer.* Transverse section through a trilobed mesenterial filament. × 320.

3. *Actinotryx Sancti-Thomæ.* Vertical section through the upper region of the column-wall, showing the mesoglœal plaitings for the support of the sphincter muscle. × 250.

4. *Actinotryx Sancti-Thomæ.* Transverse section through a mesenterial filament, just below the stomodæal region. One of the large nematocysts is represented in longitudinal section and another in transverse section. × 250.

5. *Ricordea florida.* Plan of tentacular arrangement. The members of the outermost cycle are in the same radial areas as the inner tentacles, and alternate with those of the next cycle within. The arrangement in orders is not very regular in larger polyps.

6. *Ricordea florida.* Transverse section through the stomodæal region to show the irregular arrangement of the mesenteries, and the columnar and stomodæal mesoglœal processes. Only a few of the columnar processes are represented. × 12.

7. *Stoichactis helianthus.* Plan of tentacular arrangement in a young specimen. The members of the outermost cycle communicate with the exocœles and alternate with all the radial rows, which are endocœlic.

8. *Actinoporus elegans.* Plan of tentacular arrangement.

Plate XI.

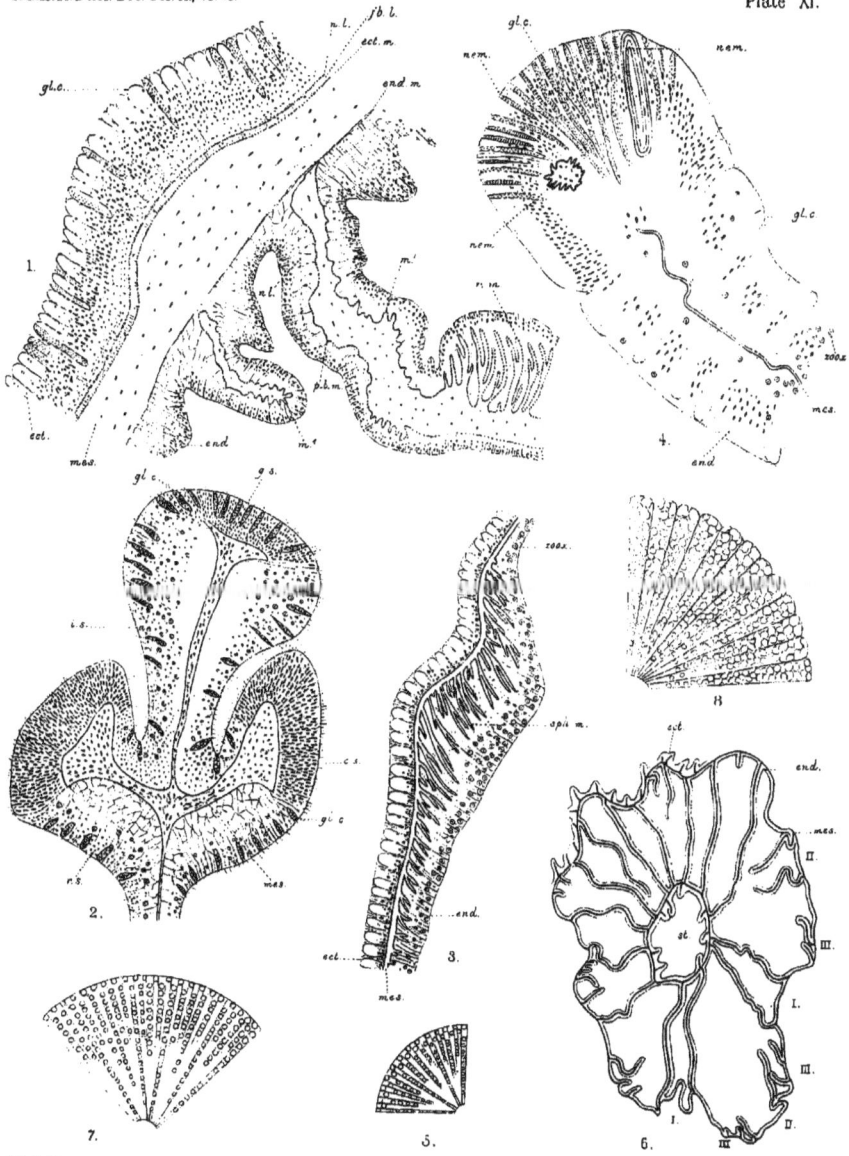

EXPLANATION OF PLATE XII.

PLATE XII.

Figure

1. *Ricordea florida.* Transverse section through the stomodœal region of a polyp in which the mesoglœal plaitings for the support of the retractor muscle are displayed, as also the irregular development of the pairs of mesenteries of the third order. Three pairs of perfect mesenteries occur on the one side of the directives and four pairs on the other. × 12.

2. *Ricordea florida.* Transverse section through a portion of the column-wall and a mesentery of the first order. × 320.

3. *Actinotryx Sancti-Thomæ.* Transverse section through a fertile mesentery with ripe spermaria in act of dehiscing. × 75.

4. *Homostichanthus anemone.* Plan of tentacular arrangement. An outer series of cycles in which the same number of tentacles occurs in each endocœle and exocœle can be distinguished from an inner series in which the cycles are imperfect, and more separated one from the other.

5. *Homostichanthus anemone.* Radial section of portion of column-wall. × 320.

6. *Homostichanthus anemone.* Tangential section through the periphery of the disc and uppermost region of the column-wall. The portion represented is but slightly tangential, so that only two mesenteries are cut through, and two tentacles are in communication with an exocœle and two with an endocœle. The mesoglœal plaitings supporting the restricted sphincter muscle are a little longer in a truly radial section than are here represented. × 25.

7. *Corynactis myrcia.* Plan of tentacular arrangement.

[214]

Trans. R. Dubl. Soc. Ser. II, Vol. VI.

Plate XII.

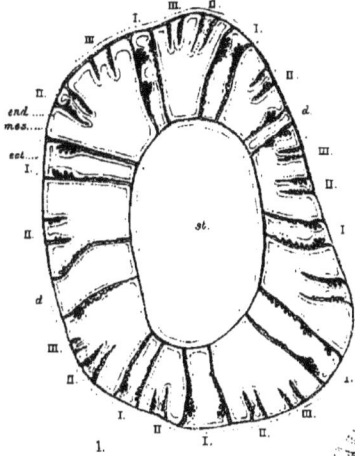

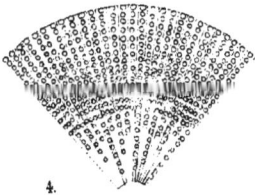

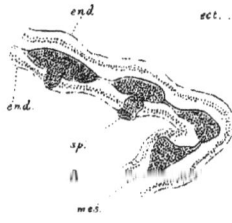

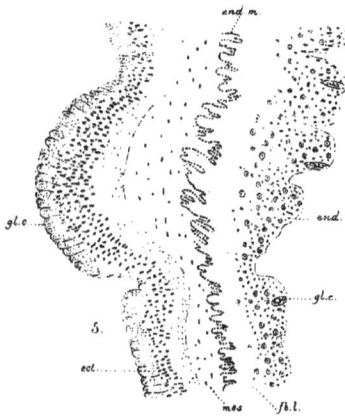

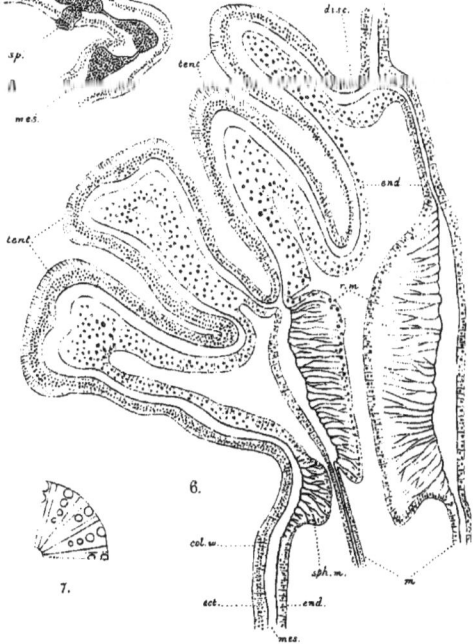

EXPLANATION OF PLATE XIII.

2 H 2

PLATE XIII.

Figure

1. *Ricordea florida.* Transverse section towards the free edge of a mesentery a little below the stomodæal region, showing the imperfect mesenterial filament. × 250.

2. *Actinoporus elegans.* Transverse section through a portion of the column-wall and of a mesentery, in the upper region of the polyp. × 320.

3. *Corynactis myrcia.* Radial section through the infolded region of the column-wall. × 250.

4. *Corynactis myrcia.* Transverse section through a portion of a polyp near the aboral termination of the stomodæum. The stomodæal wall is folded backwardly and outwardly upon itself, so that it is here cut through twice, and the enclosed endoderm is shown on the left side. The stomodæum first terminates opposite the pair of directives. As the mesenteries cease their connexion they retain around their free edge a tissue like that of the stomodæal ectoderm which passes insensibly into the mesenterial filaments. × 50.

5. *Corynactis myrcia.* Transverse section through the stem of a tentacle showing the brush-like character of the fibrillæ radiating from the mesoglœal folds. × 320.

6. *Actinoporus elegans.* Radial section through the upper part of the polyp. Slightly reduced.

7. *Parazoanthus tunicans.* Radial section through the distal region of the column-wall. × 75.

8. *Parazoanthus separatus.* Vertical section through the base. × 250.

9. *Parazoanthus monostichus.* Transverse section of a polyp through the elevated peristome and capitular region. The peripheral tissues are disorganized as a result of the presence of the numerous circularly-arranged sponge spicules. × 75.

Plate XIII.

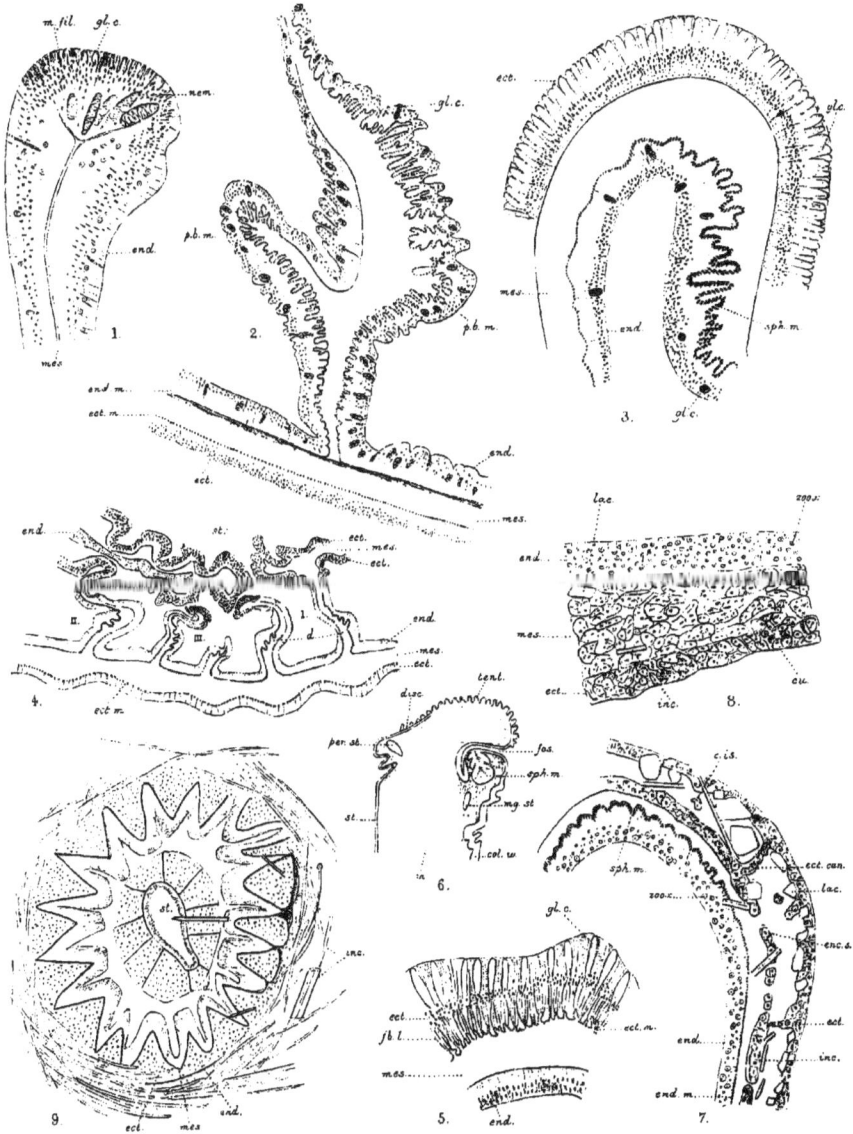

EXPLANATION OF PLATE XIV.

PLATE XIV.

LETTERING ON THE FIGURES.

col. w.,	column wall.		*m.,*	mesentery.
c. is.,	cell-islet.		*m. fil.,*	mesenterial filament.
c. s.,	ciliated streak (Flimmerstreif).		*mes.,*	mesogloea.
cu.,	cuticle.		*mg. st.,*	marginal stoma.
d.,	directives.		*mr. m.,*	macrocnemic mesentery.
disc,	disc.		*nem.,*	nematocyst.
ect.,	ectoderm.		*nr. l.,*	nerve layer.
ect. can.,	ectodermal canal.		*ob. m.,*	oblique muscle.
ect. m.,	ectodermal muscle layer.		*p. b. m.,*	parieto-basilar muscle.
enc. s.,	encircling sinus.		*per. st.,*	perioral stoma.
end.,	endoderm.		*r. ect.,*	reflected ectoderm.
end. m.,	endodermal muscle layer.		*r. m.,*	retractor muscle.
fb. l.,	fibrillar layer.		*r. s.,*	reticular streak.
fos.,	fossa.		*s. d.,*	sulcar directives.
gl. c.,	gland cell.		*sl. d.,*	sulcular directives.
g. gr.	gonidial groove.		*sp.,*	spermarium.
g. s.,	glandular streak.		*sph. m.,*	sphincter muscle.
	(Drüsenstreif, Nesseldrüsenstreif).		*st.,*	stomodæum.
inc.,	incrustations.		*tent.,*	tentacle.
i. s.,	intermediate streak.		*zoox.,*	zooxanthellæ.
lac.,	lacunæ.		*I., II., III., &c.,*	orders of mesenteries or tentacles.

Figure

1. *Stoichactis helianthus.* Radial section through the fossa and across the sphincter muscle. × 50.

2. *Homostichanthus anemone.* Transverse section through the stomodæal wall enclosing a gonidial groove. × 50.

3. *Actinoporus elegans.* Vertical section through the apex of the column and the periphery of the disc. × 50.

4. *Parazoanthus separatus.* Radial section through one-half of a retracted polyp. The section passes through a mesenteric chamber. × 75.

Plate XIV.

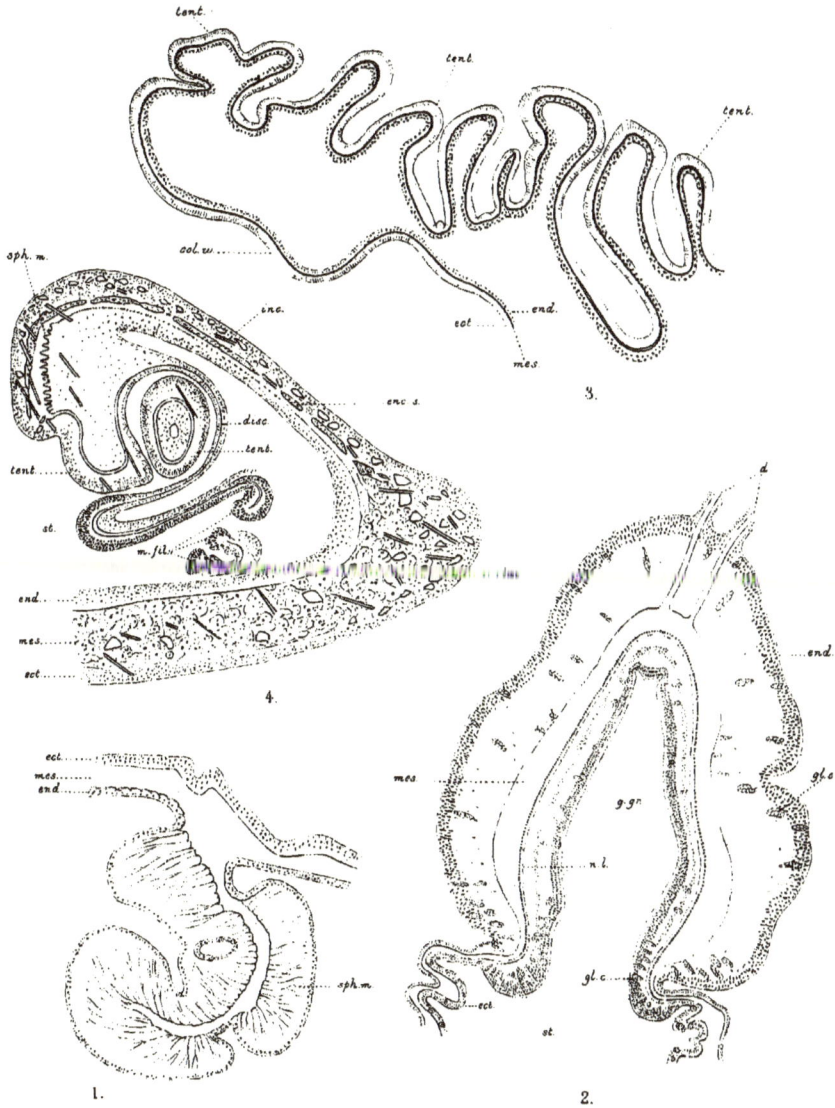

EXPLANATION OF PLATE XV.

PLATE XV.

Figure

1. *Homostichanthus anemone.* Transverse section through the knob of a tentacle. × 320.

2. *Actinoporus elegans.* Radial section through the wall of the fossa and across the sphincter muscle. × 25.

3. *Corynactis myrcia.* Transverse section through a mesentery a little below the stomodœal region. The mesenterial filament at the free edge is simple, while the endoderm behind is swollen, giving somewhat the appearance of a trilobed filament. × 320.

4. *Parazoanthus tunicans.* Transverse section of a polyp through the lower part of the stomodæum. The lacunæ represent the spaces from which the calcareous incrustations have been dissolved; the sponge spicules remain. × 50.

5. *Parazoanthus tunicans.* Section through a fold of a mesentery containing three spermaria. × 320.

Trans. R. Dubl. Soc. Ser II.Vol. VI.

Plate XV.

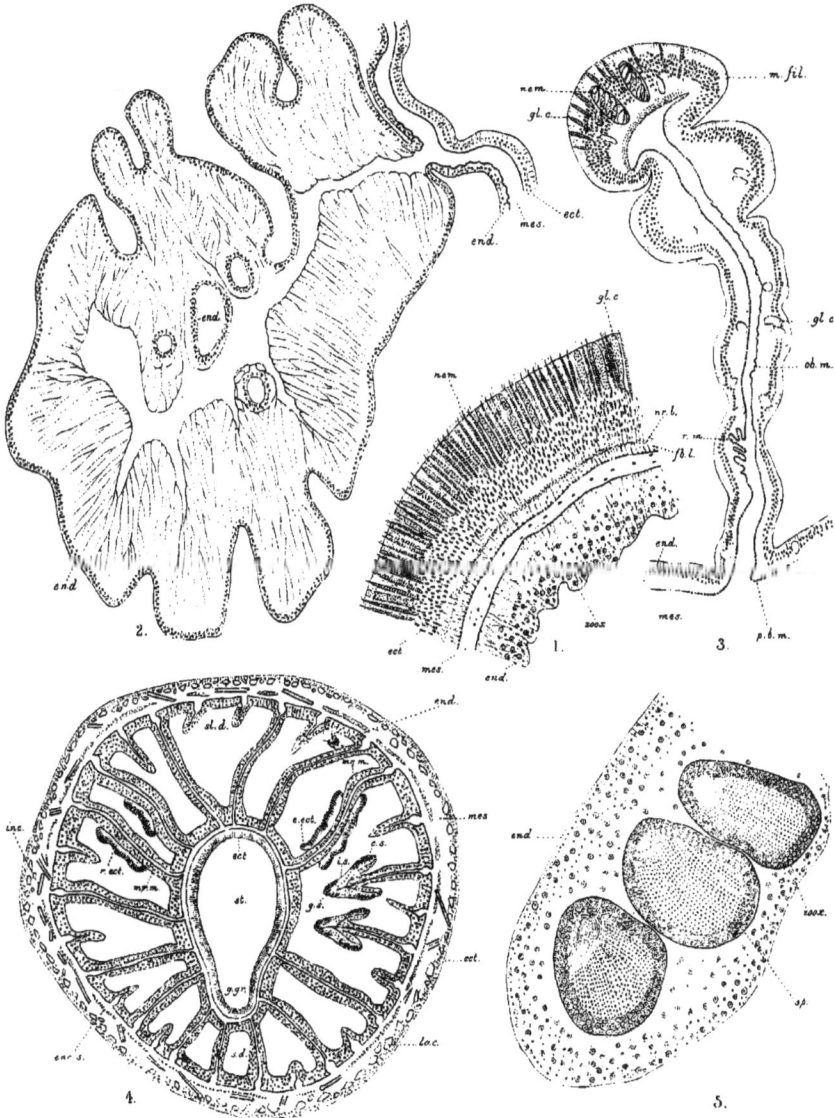

TRANSACTIONS (SERIES II.).

VOLUME VII.